Happy Apocalypse

Happy Apocalypse

A History of Technological Risk

Jean-Baptiste Fressoz

Translated by David Broder

VERSO
London • New York

This English-language edition published by Verso 2024
First published in French as *L'apocalypse joyeuse: Une histoire du risque technologique*,
Éditions du Seuil, Paris, France 2012
© Éditions du Seuil 2012
© Jean-Baptiste Fressoz, 2012, 2024
English-language translation © David Broder 2024
This translation received financial support from Labex TEPSIS.

13 5 7 9 10 8 6 4 2

Verso
UK: 6 Meard Street, London W1F 0EG
US: 388 Atlantic Avenue, Brooklyn, NY 11217
versobooks.com

Verso is the imprint of New Left Books

ISBN-13: 978-1-83976-550-6
ISBN-13: 978-1-83976-552-0 (US EBK)
ISBN-13: 978-1-83976-551-3 (UK EBK)

British Library Cataloguing in Publication Data
A catalogue record for this book is available from the British Library

Library of Congress Cataloging-in-Publication Data

Names: Fressoz, Jean-Baptiste, author. | Broder, David, translator.
Title: Happy apocalypse : a history of technological risk / Jean-Baptiste
    Fressoz ; translated by David Broder.
Other titles: Apocalypse joyeuse. English
Description: London; New York: Verso, 2024. | Translation of: Apocalypse
    joyeuse. | Includes bibliographical references and index.
Identifiers: LCCN 2023057040 (print) | LCCN 2023057041 (ebook) | ISBN
    9781839765506 (hardback) | ISBN 9781839765520 (US EBK)
Subjects: LCSH: Technology--Social aspects--History. | Technology--Risk
    assessment--History. | Industrialization--Environmental
    aspects--History. | Environmental risk assessment--History. | Technology
    and civilization. | Technological innovations--Environmental
    aspects--History. | Technological innovations--Social aspects--History.
Classification: LCC T14.5 .F73813 2024 (print) | LCC T14.5 (ebook) | DDC
    303.48/3--dc23/eng/20240203
LC record available at https://lccn.loc.gov/2023057040
LC ebook record available at https://lccn.loc.gov/2023057041

Typeset in Minion Pro by MJ & N Gavan, Truro, Cornwall
Printed in the UK by CPI Group (UK) Ltd, Croydon, CR0 4YY
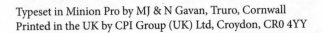

Contents

List of Abbreviations

AAM: Archives of the Académie de médecine
AAS: Archives of the Académie des sciences
AD: Archives départementales
AHPML: *Annales d'hygiène publique et de médecine légale*
AM: Archives municipales
AN: Archives nationales
APHP: Assistance publique-hôpitaux de Paris
APP: Archives of the Préfecture de police de Paris
ARS: Archives of the Royal Society (UK)
ASM: Undeposited archives of the Compagnie des salins du Midi, Aigues-
    Mortes
BM: Bibliothèque municipal [local library]
BNF: Bibliothèque nationale de France [French National Library]
BSEIN: *Bulletin de la Société d'encouragement pour l'industrie nationale*
CCI: Chamber of Industry and Commerce
Ms: Manuscript
RCS: Reports from the Conseil de salubrité (hygiene board)

Introduction: Little Modern Disinhibitions

This book examines the historical roots of the contemporary environmental crisis. It investigates the past record of technological action – how its intervention was thought about, questioned, regulated and, most importantly, imposed as the only legitimate form of life. This study dissects the forms of power, the subtle twists of reality and the moral dispositions which, at the turn of the eighteenth and nineteenth centuries, started us down the road towards the abyss. It shows that the so-called 'century of progress' was never simply technophile. The history that this book tells is not that of humanity becoming conscious of the risks of technology; rather, it is the history of a certain kind of modernising *un*consciousness, produced by political and scientific means.

In April 1855, as crowds flocked to the Universal Exhibition in Paris to admire machines, locomotives and inventions, there appeared a little book with an enigmatic title: *La Fin du monde par la science* (The End of the World through Science). Written by a lawyer called Eugène Huzar, its originality lay in the fact that this was the first critique of progress based on a catastrophist vision of the future.[1] Huzar was neither a romantic critic of the ugliness of the industrial world, nor a reactionary damning the hedonism of his times. No, Huzar loved his century, its science, and its new techniques. But he posed the problem of what *type of progress* was desirable: 'I am not waging war on science or progress, but I am the implacable enemy of an ignorant, *unprescient* science, of a progress that proceeds blindly without criterion or compass.'[2] According to Huzar,

---

1 Jean-Baptiste Fressoz, 'Eugène Huzar et la genèse de la société du risque', in Eugène Huzar, *La Fin du monde par la science*, Paris: Ère, 2008.

2 Ibid., p. 138.

since experimental science is *a posteriori* knowledge, it cannot foresee the long-term consequences of its ever more powerful products. For Huzar, this gap between technical capacities and humanity's foresight would lead inexorably to the apocalypse.

Huzar had a swarming, apocalyptic imagination: who could tell whether extracting tonne after tonne of coal risked shifting the Earth's centre of gravity and tilting its axis? Who knew whether inter-oceanic canals would disrupt sea currents, causing devastating floods? Who could predict whether vaccination would bring even more serious ills than the ones it claimed to fight? Who was to say that the carbon dioxide produced by industry would not lead to a catastrophe?

> In the course of one or two hundred years, a world now crisscrossed with railways and steam ships and covered with mills and factories would release billions of cubic metres of carbon dioxide and carbon monoxide. And since the forests were also destroyed, these hundreds of billions of cubic metres of carbon dioxide and carbon monoxide could do much to disturb the world's harmony.[3]

In their embrace of industrial development, humans believed that at most they were leaving scratches on the Earth – not realising that, according to the law of small causes and large effects, these scratches could kill it entirely. To delay the final catastrophe, Huzar proposed the establishment of a 'planetary edility': a global scientific government responsible for studying major construction projects, deforestation and all scientific experiments that could 'disturb the world's harmony'. As five million visitors shuffled through the Palais de l'industrie, Huzar's aim was clearly to provoke. In this sense, he succeeded: all the major journals published reviews of his book, most of them glowing. Writers such as Alphonse de Lamartine, Charles Dickens, Gustave Flaubert and Jules Verne read *La Fin du monde par la science* and cited it in their writings.

If Huzar was not the despised Cassandra he claimed to be, if his work was in fact widely read and debated, this was because it captured essential aspects of the techno-scientific revolution of the early nineteenth century. Huzar anchored his thinking in the controversies raised by the technologies of his time: deforestation and its climatic consequences, vaccination and the degeneration of the human species, the chemical industry and the transformation of the atmosphere, the railways and near-random

---

3   Ibid., p. 106.

disasters. Huzar's originality lay in the fact that he brought these debates together into one vast narrative, allowing him to present what his contemporaries called *progress* in a new and different way. While his theoretical approach was doubtless original, each of the arguments he used was based on debates that were already well known in the nineteenth century.

If the arguments of *La Fin du monde par la science* seem surprising, this reflects our unfamiliarity with the techno-sciences of the past and the controversies to which they gave rise. While we might imagine that proponents of a positivist modernity raved unanimously about new techniques while ignoring their long-term effects, Huzar's work offers ample evidence that such a vision was already considered retrograde during the Industrial Revolution. The men who brought about this revolution and lived through it were well aware of the huge risks they were taking. But they wilfully decided to ignore them.

The sociologists and philosophers who have reflected on environmental crisis and risk in the last four decades have run into a series of false oppositions between *modernity* and *reflexive modernity*, the society of progress and the 'society of risk' (Anthony Giddens, Ulrich Beck). Out of their concern to emphasise our own capacity to reflect on technological choices, they present a uniformly technophile past. It as if we today were the first to detect the shadow of danger amid the dazzling lights of science.

The existential stakes of the environmental crisis encouraged them to write a grandiose and rather idealist narrative of modern blindness and postmodern enlightenment. The ecological crisis is said to be the legacy of modernity itself: first, Greek science, which conceives of nature as subject to laws external to divine intentions; second, Christianity, which invented the singularity of mankind relative to the rest of creation; and finally, the scientific revolution, which replaced an organicist vision of nature with that of an inert mechanism that has to be dominated.[4]

---

4   Lynn White Jr, 'The historical roots of our ecologic crisis', *Science*, 155, 1967, pp. 1203–7; Carolyn Merchant, *The Death of Nature: Women, Ecology and the Scientific Revolution*, San Francisco: Harper, 1980; Philippe Descola, *Par-delà nature et culture*, Paris: Gallimard, 2005 (in English: *Beyond Nature and Culture*, Chicago: University of Chicago Press, 2013). According to Bruno Latour, the seventeenth-century development of an experimental community with a monopoly on the definition of facts led to a 'great divide' between nature and politics, obscuring the reality that technology was entangling these two orders (Bruno Latour, *Nous n'avons jamais été modernes*, Paris: La Découverte, 1991; in English, *We Have Never Been Modern*, Cambridge, MA: Harvard University Press, 1993).

After three centuries of frenetic modernism, transforming the world and ignoring the environment, crisis has finally arrived. The moment of reckoning has come: we, the 'postmoderns', have at last opened our eyes to the consequences of modernity. In this narrative, the notion of *risk* occupies an important place as it captures the 'reflexivity' or involution of modernity which is now confronted with its own creations. If we follow the German sociologist Ulrich Beck, risks have undergone a sudden metamorphosis: they are no longer natural, but products of modernisation; they are no longer localised, but have evolved into global uncertainties; they are no longer mere side-effects of progress, but represent the foremost challenge facing our societies. So, it is said that modernity has become *reflexive* – in other words, it is now questioning its own dynamics.[5]

The problem with this narrative is not so much that it is false, as that it is so unspecific. Its claim to reveal the deep sources of our ills makes it both intellectually seductive and politically inoffensive. Since it calls everything into question, it strikes at no actual targets. The anthropological categories on which it relies lack any political teeth – indeed, they conceal the forms of production, power and thought that, at the turn of the nineteenth century, led us down the road to the abyss. To build a serious political ecology, it is essential not to confuse the logic of the environmental crisis with that of modernity itself.

As for the idea that we are only today seeing an ecological awakening, this, too, produces a political impasse. Because it erases the reflexivity that existed in past societies, it depoliticises the long history of environmental degradation. Rather, in its emphasis on our contemporary awareness, it tends to naturalise ecological concern and overlook the social conflicts that lie behind it. By presenting a largely hypothetical reflexivity as a latter-day norm, the 'risk society' thesis has blurred the connections between past and present, and replaced historical analysis with abstract typologies. This book thus aims to combat such an abstraction of the past.

As historians pay greater attention to these questions, they are finding that modernity never had a one-sidedly mechanistic vision of the world, and never shared a unanimous project of technical mastery. Rather, a variety of cosmologies emerged, in which mastering nature did not mean contempt for it, but an understanding of its laws and a willingness to submit to them in order to act in an efficient and sustainable way.

---

5   Ulrich Beck, *Risk Society: Towards a New Modernity*, London: SAGE, 1992; Anthony Giddens, *Consequences of Modernity*, Oxford: Blackwell, 1990. For a critique, see Jean-Baptiste Fressoz, 'The Lessons of Disasters: A Historical Critique of Postmodern Optimism', laviedesidees.fr, 27 May 2011.

For example, the notion of climate is essential for understanding modern societies' reflexivity. During the sixteenth century, the concept of climate gained a certain plasticity: while it remained partly determined by people's respective geographical positions, natural philosophers became interested in its local variations, its transformations and the role of human action in its improvement or degradation. Since the climate retained its capacity to determine human and political constitutions, it became the epistemic terrain on which the consequences of technical action on the environment were thought. The determinants of health and social organisation were no longer simply a matter of one's place on the planet, but of seemingly ordinary realities (the atmosphere, forests, urban forms, and so on) that it was possible to influence in one way or another.[6]

Historians have also shown the importance of the environmental thinking which emerged from the discipline of chemistry, concerned with the exchange of matter between human society and nature. The nineteenth century was coloured by deep concerns over the metabolic rift between town and country: urbanisation – in other words the concentration of people and their excrement – was preventing mineral substances essential to the fertility of the earth from being returned to the land. All the great materialist thinkers, from Justus von Liebig to Karl Marx, as well as agronomists, hygienists and chemists, warned of the dangers of soil exhaustion and urban pollution. In the third volume of *Capital*, Marx criticised the environmental consequences of capitalist agriculture's great spaces, empty of human beings, which broke the material circulation between society and nature. According to Marx, there could be no 'getting away' from nature: whatever the mode of production, society remained dependent on a historically determinate metabolic regime; what was peculiar to the capitalist metabolism was its unsustainable character.[7]

There is no reason to look down on these theories as a kind of proto-environmentalism that would prefigure our current concern for the environment. After all, they determined modes of production that were far more respectful of the environment than our own.

For example, historians are also beginning to understand the fundamental importance of recycling in the past. *Chiffonnage*, that is, the

6   Richard Grove, *Green Imperialism, Colonial Expansion, Tropical Island Edens and the Origins of Environmentalism, 1600–1800*, Cambridge: Cambridge University Press, 1995; Jean-Baptiste Fressoz and Fabien Locher, *Chaos in the Heavens: The Forgotten History of Climate Change*, London: Verso Books, 2023.

7   John Bellamy Foster, *Marx's Ecology: Materialism and Nature*, New York: Monthly Review Press, 2000.

collection of abandoned objects and materials, employed nearly 100,000 people in 1860s France. Bones, rags, metals – everything was resold and reused. Up until the end of the nineteenth century, urban *excreta* were systematically recycled for agricultural purposes.[8] The city police of the Ancien Régime provide another example. Since their work was rooted in neo-Hippocratic medical notions that made 'airs, waters, and places' the determinants of health, they paid close attention to alterations in the urban environment. Similarly, the regulations on forests and fisheries (Jean-Baptiste Colbert's statute on water and forests, the Marine Ordinance of 1681) remind us of the state's involvement in conserving resources and the severe penalties (fines, imprisonment and corporal punishment) that were attached to these environmental rules. Finally, the existence of 'commons without tragedy' – the significant fact that communities managed to preserve natural resources (fisheries, forests, pastures) for centuries – bears witness to the ecological intelligence of past societies.[9]

So, when it comes to writing history, it is misleading to tell a story of the Industrial Revolution in which societies unconsciously altered their environments and ways of life, and only later understood the dangers this brought and the mistakes they had made. Past societies did not cause environmental havoc inadvertently, or without contemplating the consequences of their decisions. Confidence in the future could not be taken for granted, and, at each strategic point in modernity, at each point of conflict, ignorance and/or disinhibiting knowledge had to be produced in a calculated way.

Considered in terms of their effects, the *disinhibitions*, whose history this book follows, generally include all sorts of devices that made possible, acceptable and even desirable the technical transformation of bodies, environments or modes of production. For any innovation of real significance to establish itself, it has to overcome moral reticence, social opposition, vested interests, suspicions, and criticisms of its material and environmental consequences. The confidence that presides over the technical transformation of the world requires theories that blur its meaning and cushion its traumatic impact. After disasters, there is a need for discourses and moral dispositions that will neutralise their effects, playing down their ethical dimensions in order to ensure the continuation of the technological project.

---

8   Sabine Barles, *L'Invention des déchets urbains. France 1790–1970*, Seyssel: Champ Vallon, 2007.

9   Elinor Ostrom, *Governing the Commons: The Evolution of Institutions for Collective Action*, Cambridge: Cambridge University Press, 1990.

The word *disinhibition* encapsulates the two distinct phases of technological action: reflexivity and disregard, contemplating danger and normalising it. Modernity was a process of reflexive disinhibition: we shall see how the regulations, consultations, safety standards, authorisation procedures and health surveys that purported to understand and contain risk generally had the effect of legitimising technological *faits accomplis*.

Finally, the insistence on the smallness and ad hoc nature of these disinhibitions is intended to point out that modernity is not the majestic and spiritual movement that philosophers claim it is. Rather, I would like to think of it as a mass of little *coups de force*, imposed situations and normalised exceptions. Modernity was an endeavour that had to be brought to fruition. Those who made this happen produced knowledge and ignorance, legal norms and discourses which aimed to establish new sensibilities and new ways of conceiving our lives, our bodies and our relations with environments and objects.

This book also takes a fresh look at the so-called Industrial Revolution. Historians concerned with the causes of European (or rather, British) exceptionalism have related it to various historical events. It is connected to a transformation in industry or to a slow transformation of agriculture and artisanal labour, to the existence of efficient markets, to the protection of private and intellectual property, to a growth in demand and an intensification of work, to the exploitation of new sources of energy or to the ecological assistance drawn from the New World.[10] Through these preoccupations, the technological dimension of the transformations that took place between 1750 and 1850 has – for the most part appropriately – left the historiographical centre stage. This book turns the focus back to innovation – not to make it the decisive cause, but to pose it as a problem. How was the acceptance of technology achieved? How were technophilic subjects and new understandings of reality created, such as could make the world compatible with the needs of the Industrial Revolution?

When the history of innovation ceases to naturalise its object of study, it can also be politically liberating. Instead of reducing opposition to a simple resistance, imagined to be self-interested or irrational and unable to propose any alternative, it can, instead, show the often indecisive and

---

10   For a recent overview, see Kenneth Pomeranz, *The Great Divergence: China, Europe and the Making of the Modern World Economy*, Princeton, NJ: Princeton University Press, 2000, pp. 29–68; Jan de Vries, *The Industrious Revolution: Consumer Behavior and the Household Economy, 1650 to the Present*, Cambridge: Cambridge University Press, 2008.

sometimes even contingent nature of past technological choices. The first approach I took in this book was to reconsider the technologies emblematic of the medical and industrial revolutions – inoculation and vaccination against smallpox, the chemical industry and gas lighting, steam and rail technology – as *cases*, in the casuistic and moral sense of the word. I wanted to show that these technologies were, in their own time, objects of doubt, dispute, misgivings and perplexity, much as contemporary technoscience can be. To give a sense of the original indeterminacy of these controversies, I have taken seriously each of the players therein – all of them, even those who may seem most obviously mistaken. This approach, which has the merit of giving a voice back to the losers of history, is also a heuristic choice. For making sense of out-of-the-ordinary arguments also forces us to reconstruct frameworks of intelligibility that their defeat rendered invisible. In so doing, it becomes clear that the opponents of these projects were not siding *against* innovation, but rather *for* their environment, their safety, their jobs, and for the preservation of forms of life they considered valuable.

When we get into the complexities of these disputes, the notions of innovation and resistance become rather blurred, and our concept of a single line of technological time goes haywire.[11] For example, in 1819, in the early days of gas lighting, the chemist Nicolas Clément-Desormes explained that the axis of time punctuated by innovations was not an axis of their steadily rising value. Gas lighting, with its profusion of underground pipes, was merely a costly imitation of the marvellous simplicity of the oil lamp, which also had the merit of being portable, of using renewable resources and preserving individual autonomy. Clément-Desormes turned on its head the very notion of innovation: 'Let's suppose that gas lighting was the first known [technology], that it was in use everywhere, and that some man of genius presented us with an Argand lamp or a simple lit candle. What admiration we would have for such an astonishing simplification.'[12] In fact, rather than discussing innovation and resistance, we would better speak of the competition between different technological projects: the choice for improved oil lamps, rather than gas lighting; systems of animal traction, rather than the steam engine; quarantines and hygiene measures rather than inoculation and vaccination; plant-based materials rather than chemicals; and so on.

---

11   David Edgerton, *The Shock of the Old: Technology and Global History since 1900*, London: Profile Books, 2006.

12   Nicolas Clément-Desormes, *Appréciation du procédé d'éclairage par le gaz hydrogène du charbon de terre*, Paris: Delaunay, 1819, pp. 33–41.

Despite everything, the techniques which I have studied did eventually come to the fore (albeit, on a mass scale, only towards the end of the nineteenth century). The advantage of going off track and focusing on the indeterminacy of these disputes is that, by relativising the intrinsic superiority of the innovations in question, we are able to consider power and the means by which it operates. This book explains in some detail the forces that led to the victory of these technical systems, despite their dangers, despite the opposition they faced, and despite *awareness* of their dangers. I have sought to understand why the technologies that have produced our modern world and contemporary environmental crisis came into being – for whom and against whom, based on what knowledge and despite which other knowledges.

This book's second goal is to re-examine the question of risk in terms of paths of technological development. The emergence of the precautionary principle at the end of the last century and the hope for a democratic governance of techno-sciences have led sociologists to pose the problem of risk within a decisional framework. In this view, the crucial question would be that of the choice of technology, the assessment of risks and the (democratic and transparent) procedures that need to be established to guide this choice and these evaluations.[13] The problem is that, historically, technology has never been the subject of a shared choice. Some actors have actively brought it into being, and then it has had to be regulated. Contrary to the sociological dream of a controlled technoscience, the history of technology is one of innovations being forced through, and the subsequent efforts to normalise these *faits accomplis*. By taking a long-term perspective, the question of risk can be grasped in a completely different way. The relevant question is no longer so much about choice and the right processes for making the correct choice, but, rather, the path of technological development and the various ways of influencing it. I thus propose to take advantage of the historian's ability to consider forms of technology in a long-term perspective. This allows me to write a comparative history of the different ways in which risk has been regulated (by technological standards, by recourse to the courts, by administrative oversight, by insurance) and their effects on knowledge and the paths of technological development.

13   Michel Callon, Pierre Lascoumes and Yannick Barthe, *Agir dans un monde incertain, essai sur la démocratie technique*, Paris: Seuil, 2001.

The first chapter looks at the famous controversy around smallpox inoculation to describe the historical context in which risk emerged.[14] Risk, understood here as the application of probabilities to human life, arose in the 1720s to persuade individuals to get inoculated. In particular, I explain the theological and political frameworks that justified this extension of the geometry of chance to life itself. But the question of inoculation remained inextricably moral, religious, political and corporeal, and statistical arguments failed to produce a quorum of brave subjects prepared to risk their lives in order to preserve them.

In contrast, the case of smallpox vaccination, the subject of the next chapter, provides an opportunity to study the techniques of proof that ensured the success of the imperial policy of the 1800s. Napoleon's administration, which saw the public as incapable of devising risk-minimisation strategies, set out to demonstrate that the vaccine was perfectly safe. Human experimentation, the clinic, the graphic definition of disease and the statistical management of medical information made it possible to impose the improbable definition of a non-virulent virus, a *perfectly benign* virus that would keep smallpox at bay *forever*. The government intervened in a reasoned way in the medical field so as to disseminate statements which it deemed strategically important for the national good. In this case, modern disinhibition was indeed a form of manipulation and imposition of perceptions, seeking to capture, mobilise and align behaviour as the new techniques required. Here science was primarily a form of disciplinary enterprise, and not so much a method for exploring reality.

The following two chapters are also about bodies, but, this time, in connection to the environment. Contrary to common assumptions, the environment is not at all a recent preoccupation. It has a long genealogy, part of which I characterise through the notion of *circumfusa* ('surrounding things' in Latin). This category of eighteenth-century hygiene had an extremely broad reach: it included all natural and artificial entities (waters, climate, artisanal odours and smokes . . .) that determined the health and even the shape of bodies. The paradox, then, is that industrialisation and the alteration of surroundings that it brought with its trail of pollution,

---

14   Historians estimate that, in the eighteenth century, one in seven people in Europe died of smallpox, usually in infancy. Smallpox inoculation was the insertion of the pus produced by this disease in the hope of catching a benign form and becoming immune to it. This technique, probably known since the eleventh century in China, appeared in Europe in the 1720s. *Vaccinia* is a cow disease discovered by the English physician Edward Jenner in 1798. Vaccination involves injecting the pus from this disease to protect against smallpox.

took place not in an intellectual void, but despite the dominance of medical theories that considered the environment as the decisive force shaping human's health and constitution.

In France, the 1800s were once again crucial. We will see how chemical capitalism presided over a radical transformation in environmental regulation. Instead of the urban police of the Ancien Régime, who had been concerned with health protection and could readily ban or punish artisans, the Napoleonic administration set up a liberal framework for regulating conflict, based on simple financial compensation for environmental damage. At the same time that environments were being commodified, hygienism was reshaping medical theories: it was now *social* factors that determined the health of populations, rather than environmental ones. So, in increasing society's wealth, industrialisation was meant ultimately to produce a healthier population.

The last two chapters are devoted to industrial risk and its management. Contrary to sociologists' assumptions, the risks of the industrial age were immediately major ones. It has never been possible to simply externalise risk and delegate its management to insurance companies. Risk has always been a political question of spatial and social distribution. Through a case study of the beginnings of gas lighting in Paris, I show that the safety norm – that is, the new project to make the world of production safe by defining the right technological forms – appeared in order to legalise the risks of gas, despite the complaints of city dwellers. Such norms were also essential in order to maintain the principle of individual responsibility in a technological society: they were supposed to bring about perfect technologies whose failures would be entirely attributable to human error.

Let us conclude this introduction with an anecdote that shows the importance of historical narrative in our current understanding of innovation. Everyone knows, or thinks they know, the famous story of how our ancestors were afraid of trains. This fable is always cited in order to discredit irrational fears about science. In 2004, at the height of the GMO controversy in France, the CEO of a biotech start-up explained in the pages of *Le Monde* that 'the countless articles written to terrify public opinion could fill a catalogue of howlers, of the same level as what was written at the time the railroads first appeared'.[15] The previous year, the philosopher of science Dominique Lecourt denounced 'biocatastrophists' by referring

15   Michel Debrand, 'Sauvons les OGM', *Le Monde*, 8 September 2004.

to the irrational fears spurred by the first railways: 'In 1835, the members of the Lyon academy of medicine solemnly asked: won't we risk damage to the retina and breathing problems at high speed, and won't pregnant women being tossed around suffer miscarriages?'[16] History makes it possible to ridicule anxieties through a kind of technophile induction: just as past fears about well-established technologies seem absurd to us today, our present fears about innovation will seem ridiculous to our descendants. The importance that this argument has assumed is worth dwelling on.

In fact, it is a myth. In 1863, Louis Figuier, the great nineteenth-century science populariser, reviewed a paper by the doctor Pietra Santa on the health consequences of the railways.[17] Figuier took the opportunity to write a little compendium of fake medical nonsense. He mentioned, without giving any references, accusations which had supposedly been levelled by learned doctors: the railways were said to strain the eyesight, cause miscarriages and nervous disorders. It is true that doctors in the 1850s questioned the effects of the railways on health. Obstetrics textbooks contain brief references to the dangers of long journeys (by horse carriage or train) and excessive vibrations for pregnant women nearing term.[18] The supposed influence of the railway on eyesight is more mysterious. Pietra Santa advised readers to rest their eyesight regularly, but made no specific claims regarding the railways. Whatever Figuier's sources, there is, in any case, no trace of the report which Dominique Lecourt mentions, published by the Lyon academy of medicine – an institution which, incidentally, never existed.

The construction of this myth continued in Germany. In 1889, Heinrich von Treitschke mentioned, without giving any references, a report from 1835 (the same date as the alleged Lyon report) by the Bavarian Medical College, which recommended banning the railways because their incredible speed could cause 'delirium furiosum' in passengers. This anecdote was an extraordinary success. It appeared in a history of the railways in 1912, in *Mein Kampf* in 1925 (in which Hitler used it to ridicule experts), and then in various historical works on the Industrial Revolution, from the 1960s to the 1980s, each time mentioned in connection with 'resistance

16   Dominique Lecourt, 'Faut-il jeter le clonage avec le fantôme d'un bébé clone?', *Chronic'art*, 10, 2003.

17   Louis Figuier, 'L'Hygiène et les chemins de fer', *L'Année scientifique et industrielle*, Paris: Hachette, 1863, pp. 389–96.

18   Amédée Dechambre, *Dictionnaire encyclopédique des sciences médicales*, Paris: Masson, 1886, vol. 47, entry 'Grossesse'.

to progress'. Of course, the Bavarian report, like its counterpart from Lyon, never existed.[19]

In France, embellished by Treitschke's imagination, the story prospered. In 1906, in the preface to his *Bibliographie des chemins de fer*, Pierre-Charles Laurent de Villedeuil added, among the presumed medical consequences of train travel, 'danse de Saint-Guy' (Sydenham's chorea) produced by the vibrations. He also refers to blindness using an outdated medical term: railways were said to 'enflame the retina' because of the rapid succession of images. Yet, despite the 826 pages which this bibliographical compendium boasted, no references were offered![20] In 1957, an article in *L'Express*, marking the 120th anniversary of the opening of the Paris-Saint-Germain line, explained that in the 1840s, 'sinister oracles' had announced that the railways would 'cause nervous diseases, even epilepsy and Sydenham's chorea', and that they would 'inflame the retina and cause pregnant women to miscarry'. The author adds: 'These were not the mutterings of bonesetters, but prophecies publicly communicated to the Academy of Medicine.'[21] The story was later retold in popularising works and even university history textbooks. The myth became part of 'general knowledge'. To this day, it is still routinely invoked by politicians when there is any sort of debate around an innovation.[22]

In conclusion, we may note that in 1860, at a time when the rumour of a link between madness and railways was taking form, the medical profession and the courts were beginning to study (and compensate) the nervous trauma caused by rail accidents, which had nothing to do with Sydenham's chorea.[23] In fact, the countless complaints, lawsuits and petitions were not opposed to the railways themselves, but had to do with the accidents they caused and the companies suspected of cutting corners at the expense of passenger safety. The safety of today's rail systems is the welcome offspring of these protests.

---

19   Bernward Joerges, 'Expertise lost: an early case of technology assessment', *Social Studies of Science*, 24, 1994, pp. 96–104.

20   Pierre-Charles Laurent de Villedeuil, *Bibliographie des chemins de fer*, Paris: Librairie générale, 1906, p. 46.

21   Georges Ketman, 'Ce Jour-là', *L'Express*, 23 August 1957.

22   André Philip, *Histoire des faits économiques et sociaux de 1800 à nos jours*, Paris: Aubier, 1963, p. 94 (republished by Dalloz, 2000); Simone de Beauvoir, *La Vieillesse*, Paris: Gallimard, 1970, vol. 2, p. 416; Henri Vincenot, *La Vie quotidienne dans les chemins de fer au XIXe*, Paris: Hachette, 1975, p. 13.

23   Wolfgang Schivelbusch, *The Railway Journey: The Industrialisation of Space and Time*, Berkeley, CA: University of California Press, 1986, pp. 134–49; Ralph Harrington, 'The railway journey and the neuroses of modernity', in Richard Wrigley and George Revill (eds), *Pathologies of Travel*, Amsterdam: Rodopi, 2000, pp. 229–59.

1

Inoculated with Risk

Risk – the application of the probabilities to matters of life, death and health – is the most generally used tool to guide individual behaviour. Confronted, for example, with the question of whether or not to be vaccinated, we have to rely on simple fractions and subsume our lives and our bodies into statistics. Risk calculations tell us to behave like rational individuals seeking to maximise our life expectancy.

For a long time, the use of risk remained confined to the world of commerce and insurance. In the mid-twelfth century, Pisan and Genoese merchants borrowed the Arabic word *rizq* (the share that God assigns to each man) to refer, in their contracts, to the losses and profits associated with unpredictable events. Risk, along with account-keeping and maritime insurance, contributed to the growth of merchant capitalism. It also contributed to its moral legitimisation: commercial profit, that is, the merchant's gain from reselling an unchanged good, was justified in theological terms through reference to the 'price of risk'. In the seventeenth century, with the development of statistics (or political arithmetic) and insurance (for fire and life), risk was a tool to compensate financially for the world's dangers.[1]

But through what historical process could this concept, rooted in commerce and insurance, become the generic contemporary tool for

---

1   Sylvain Piron, 'L'apparition du *resicum* en Méditerranée occidentale, XIIe–XIIIe siècles', *Pour une histoire culturelle du risque. Genèse, évolution, actualité du concept dans les sociétés occidentales*, Strasbourg: Histoire et Anthropologie, 2004, pp. 59–76; Geoffrey Clark, *Betting on Lives: The Culture of Life Insurance in England, 1695–1775*, Manchester: Manchester University Press, 1999; Cornell Zwierlein, 'Insurances as part of human security, their timescapes and spatiality', *Historical Social Research*, vol. 35, no. 4, 2010, pp. 253–74.

governing all sorts of behaviour? Smallpox inoculation, a medical technique that appeared in Europe in the eighteenth century, provided the decisive opportunity to extend probabilistic rationality to the government of life.

Inoculation is a minimalist medical technique: some pus of variola, an incision, a few days' fever, and the hope of being finally exempt from a terrible disease. In exchange for a few pustules, one could escape smallpox – an illness that often led to death, or left one in disgrace, sometimes blind, or deaf, or debilitated, or all of these things at once. While low-tech, inoculation was, nevertheless, one of the most important innovations of its age: not only because it played a certain role in the demographic revolution, but, above all, because it heralded a new way of living with viruses. Rather than resorting to an embargo strategy (quarantine, isolation, surveillance, disinfection), humans made a deal with their nemesis: in exchange for immunity, they gave sanctuary to a few viruses that were kind enough not to harm their hosts too much.

Inoculation played a key role in the Enlightenment project of an autonomous individual guided by reason: it was the emblematic technique of a moral philosophy that valued rational self-government.[2] Those who had themselves inoculated, or who hesitated to do so, had to deal with the same doubts: was it legitimate to run the risk of death deliberately, or to make one's children do so? Worse still, as inoculated people became temporarily contagious, they forced an increased risk on others. There were two opposing positions: the inoculators defended the freedom of individuals to protect themselves; their opponents stood for the collective discipline demanded by the policing of public health. By making it possible for individuals to escape humanity's shared biological lot, inoculation raised new ethical problems. The Baconian project of a science which would 'elevate and assist mankind' necessitated some moral assistance to assuage ethical doubts.[3] For the propagandists for inoculation, risk precisely fulfilled this function.

In this chapter, I will highlight the radical strangeness of what other historians have taken for granted: namely, the application of the geometry of chance to the management of life. The aim is to give a sense of the revolution that risk represented in the 1720s: a new way of looking at life and its place in the natural order, unexpected bearings for individual

---

2   Jérôme B. Schneewind, *L'Invention de l'autonomie. Une histoire de la philosophie morale moderne*, Paris: Gallimard, 2001.

3   Francis Bacon, *Novum Organum*, 1620, New York: Collier and Son, 1903, p. 50 (aphorism 73).

action, and a novel source of legitimacy for exerting power. Retracing the history of inoculation, without considering probabilities as the *natural* way of thinking about it, sheds light on the specific historical context that led to the emergence of risk as a way of governing conduct.[4]

The concept of the risk of inoculation appeared in England and New England in the 1720s, as a casuistic tool for the resolution of moral misgivings. It was supported by a theology that linked the probabilistic order of the world to the divine order. In the 1750s, in France, risk took on a different meaning: it was part of a philosophical project that valued autonomous, rational individuals and actively sought to mobilise them for the public good. Risk was also at the heart of the utopian vision of the public sphere: it separated the essential (numbers) from the accessory (moral arguments), and distinguished legitimate judges (rational men reading statistics) from the frivolous and sentimental crowd (the people, children, mothers). Finally, from the 1760s onwards, risk was used in a political project that Michel Foucault famously called 'biopower'. The very first numerical simulation of the demographic effect of inoculation created a new viewpoint: that of a monarch wishing to optimise his population, to maximise the number of useful subjects (active and fertile ones) and minimise the number of children needed to produce them. This rationality was never put into practice under the Ancien Régime, except in the case of the slave master.

1. The casuistic emergence of risk

The first probabilistic arguments in favour of inoculation were developed in Boston in 1721. As a smallpox epidemic spread through the city, a group of pastors, led by Cotton Mather, encouraged the faithful to have themselves inoculated. They preached on the subject and published pamphlets and articles. William Douglass, the only doctor in the city to have attended medical school, sarcastically dubbed them the 'inoculation ministers'. Opposition was particularly strong: since epidemics were conceived as 'judicial diseases' punishing the sinful community, inoculation was rejected as an impious attempt to escape the divine will.

---

4   The history of quantification practices has already addressed the question of inoculation, but without highlighting the theological and political coup that the probabilistics of human life represented. See Andrea Rusnock, *Vital Accounts: Quantifying Health and Population in Eighteenth-Century England and France*, Cambridge: Cambridge University Press, 2002. On inoculation in general, see the indicative bibliography at the end of this volume.

Mather does not seem to have anticipated such a reaction. In his early writings on the subject, he presented inoculation as the natural consequence of the desire for security, as was also reflected in the development of life insurance. How could Bostonians refuse inoculation when 'many are prepared to pay a high price to insure their lives against the dangers of this terrible disease'?[5] On 14 November 1721, a bomb was thrown into his house. The device was wrapped in a note: 'Mather, You Dog, Damn you, I will enoculate you with this, with a Pox to you.'[6] In this context, pastors had better be particularly persuasive.

The main obstacle the proponents of inoculation had to overcome was the instinctive repugnance towards this practice. The disgust felt towards a virulent pus, common sense about purity and danger, even the instinct of animals who 'know how to make use of natural things' and who do not have themselves inoculated – were these not so many signs of the existence of a universal and thus divine prohibition against this practice? A sign of what St Augustine had written about: the 'incorporeal light', a good engraved by God in our hearts, 'by which our minds are somehow irradiated, so that we may judge rightly of all these things'.[7] The difficulty for inoculators was to present a new action on the body, which seemed patently unnatural, as itself part of the natural order.

Risk served precisely this purpose. In order to strike down misgivings and justify inoculation, the pastors changed the relationship between nature and morality. They did so by articulating these two elements not through the universal sentiments deposited by God in his creature, but rather through probabilistic laws, which had to be discovered in the world.

According to the logic of natural theology, since the world was ordered by Providence, only by studying its laws was it possible to understand divine decrees and act morally. In 1700, Mather published a treatise entitled *Reasonable Religion*, demonstrating that 'whoever acts reasonably lives religiously'. The moral problem posed by Mather was topographical: he wanted to shift the site of moral judgement from the subject's inner life to the exercise of a reason that grasped the world's regularities. Acting morally consisted in rejecting spontaneity and behaving in accordance

---

5   Letter from Mather to the Boston physicians, 6 June 1721, quoted in *A Vindication of the Ministers of Boston*, Boston: B. Green, 5 February 1722. Between 1696 and 1721, more than sixty life insurance schemes were founded in England. See Clark, *Betting on Lives*, p. 33.

6   Cotton Mather, *Diary of Cotton Mather (1709–1724)*, Boston: Peabody Society, 1912, vol. 2, p. 658.

7   Saint Augustine, *The City of God*, quoted in Charles Taylor, *Sources of the Self: The Making of Modern Identity*, Cambridge, MA: Harvard University Press, p. 136.

with an external, natural and divine law.[8] Before the Fall, man was able to follow his own inclinations, because divine law was 'inscribed in his heart'. After the Fall, he had to call on the reason God had given him to decipher the 'hieroglyphs' of nature. The true directors of conscience are natural beings: 'Many things, though silent, cry out just reproaches against us.'[9] The good Christian understands God's designs, which shine through from the order of the world.

The justification for inoculation proposed by the Bostonian pastors was an application of this natural theology. Mather rejected theoretical and biblical arguments either for or against inoculation: 'Experience! 'tis to Thee that the Matter must be referr'd after all.'[10] To the misgivings of a candidate for inoculation, worried about his future salvation, Pastor Colman replied: 'If someone dies from inoculation, he dies in the use of the *most probable* means he knew to save his life; he dies in the way of duty and therefore in the way of God.'[11] Moral doubts could thus be soothed by comparing the risks of natural smallpox and inoculation.

If risk was used to justify inoculation in Boston in 1721, this was because quantitative information was readily available there: unlike in Europe, smallpox was not endemic in New England, and, as soon as the epidemic began, caused by a ship entering Boston harbour, the authorities introduced strict quarantine measures. To organise these provisions, not only the dead were counted, but also those who fell ill with smallpox. The ratio of dead to sick was used to calculate the risk of natural smallpox, which was then compared with the results of the inoculations. Mather and Colman thus set out the first quantitative arguments in favour of the practice: during the Boston epidemic, out of 7,989 patients, 844 died – 1 in 9 – whereas, out of the 286 inoculated, only 6 died – or 1 in 48. The considerable and constant difference between the two mortality rates clearly testified to the existence of a divine order in favour of inoculation.

---

8   On Protestant asceticism's rejection of the inner voice, see Max Weber, *The Protestant Ethic and the Spirit of Capitalism*, London: George Allen & Unwin, 1930.

9   Cotton Mather, *Reasonable Religion: Or the Truths of the Christian Religion Demonstrated*, London: Cliff and Jackson, 1713, p. 39. On natural theology or physical theology, fundamental to natural philosophy in the early eighteenth century, see Jacqueline Lagrée, *La Religion naturelle*, Paris: PUF, 1991; Jean-Marc Rohrbasser, *Dieu, l'ordre et le nombre*, Paris: PUF, 2001; Margaret J. Osler, 'Whose ends? Teleology in early modern natural philosophy', *Osiris*, 16, 2001, pp. 151–68.

10   Cotton Mather, *An Account of the Method and Success of Inoculating the Small-Pox in Boston in New England*, London: J. Peele, 1722, p. 15.

11   Reverend Colman, *A Letter to a Friend in the Country Attempting a Solution of the Scruples and Objections of a Conscientious or Religious Nature Commonly Made against the New Way of Receiving the Small-Pox*, Boston: S. Kneeland, 1722.

The story of risk continued in England. In the 1720s, Mather and Colman exchanged letters with the physician James Jurin, secretary of the Royal Society.[12] It was Jurin who, from 1723 onwards, undertook to promote inoculation in England. Since smallpox was endemic in London, it was more difficult to quantify the danger. Jurin made use of the parish mortality bulletins, which since the early seventeenth century had detailed the causes of death (an essential resource that French inoculators lacked). He also conducted small censuses, family by family, to estimate the risk of dying from smallpox once the disease had been diagnosed.[13] Finally, he published an advertisement in the newspapers inviting doctors to send him the results of their inoculations. In 1724, a widely distributed pamphlet reported on this research.[14]

In London as in Boston, probabilistics made sense thanks to natural theology. For Jurin, the difference in risks revealed the divine origin of innovation. The reverend David Some included small arithmetical demonstrations in his sermons. In his view, to take risk of inoculation was not a sign of impiety, but, rather, a reminder of man's dependence on God.[15] The quantification of life was also part of a movement to rationalise charity. In 1752, the Bishop of Worcester founded the Smallpox and Inoculation Hospital in London, which provided free inoculations. To encourage donors' generosity, in a sermon he calculated the population that England would gain if the practice became widespread. A few years later, another minister went so far as to calculate the number of lives saved for every pound donated to the hospital.[16]

In 1754, the probabilistic argument appeared in two treatises on inoculation written in French: L'Essai apologétique sur l'inoculation by Charles

---

12   ARS, ms 245, 'Extract of a letter from the Revd M. Colman from Boston to Henry Newman', 8 March 1722; James Jurin, 'A letter to the learned Coleb Cotesworth . . . containing a comparison between the danger of the natural smallpox and of that of the given by inoculation', Philosophical Transactions, 32, 1722, p. 215, which quotes a letter by Mather.

13   ARS, ms 245, Anthony Fage, 'An account . . . of how many had the small pox in the year 1722, how many had it before . . . and how many died'.

14   James Jurin, An Account of the Success of Inoculating the Small-Pox in Great Britain, London: Peele, 1724, which was immediately translated into French. One of the original aspects of this work was its attempt to reconcile quantification and medical narrative. Most of the text is devoted to accounts of fatal inoculations, so that readers can judge the causal link between this intervention and death. Depending on the reader's assessment, the risk of death from inoculation varies considerably, from 1 in 64 to 1 in 482. For more on Jurin, see Rusnock, Vital Accounts, pp. 49–70.

15   David Some, The Case of Receiving the Small-Pox by Inoculation, 1725, London: Buckland, 1750, p. 27.

16   Samuel Squire, A Sermon Preached before His Grace Charles, Duke of Marlborough . . . March 27, 1760, London: Woodfall, 1760.

Chais; and *L'Inoculation justifiée* by Samuel Tissot. Chais, already known for his works on theology, was a pastor of Genevan origin living in The Hague. Tissot, who would become a famous medical figure, was still but a young doctor in Lausanne. *L'Inoculation justifiée* allowed him to make his name. This was not a simple medical treatise, but, as the subtitle indicates, an *apologetic dissertation*. The work was dedicated to his uncle, a pastor in a village near Lausanne.

The two authors each proposed a monetary metaphor, in order to make the equivalence between risk and divine will more tangible. It would be wrong to speak of life as a 'gift from God'. My life does not belong to me. It is merely a 'valuable deposit' that God has entrusted to me and for which 'I am accountable'. This expression ought to be taken literally: at the Last Judgement I will have to 'give account' of my decisions before the Great Owner. Not choosing the best risks is tantamount to harming God's interest.[17]

The Protestant casuistry of inoculation describes a world with no obvious natural law, no inner voice to tell us what is right. Man is 'left to his own devices . . . without an infallible oracle'. The only course to take is the one that 'seems *probably* the best . . . We can then say, Providence is calling us to it'.[18] The morality of inoculation was based on intimate conviction: the faithful would have to meditate on the difference in risks and its theological meaning, and finally recognise that the probabilistic argument is not a simple opportunism cloaked in the law of nature. Once he is convinced that probabilities point to God's way, 'he must no longer proceed in the spirit of misgivings, for success depends on the dispensations of Providence'. Once thus convinced, the Protestant can face the uncertain outcome with ardour, for he is part of the providential order; risk is simply an expression of God's grace, to which we must entrust ourselves. Protestant apologists for inoculation could reject natural feeling in the name of probabilistic reason, without being accused of putting the useful before the authorised or the physical before the moral. For probabilistic reason is, at the same time and above all, theological reason.

Inoculation and its probabilistic justification were part of a wider historical process that moved the definition of the right action from the Word of God to an understanding of the laws of nature. John Locke's natural theology and eudemonistic ethics were essential to this transition. Because nature obeys laws chosen by God, because man obeys the

---

17   Charles Chais, *Essai apologétique sur l'inoculation*, The Hague: De Hondt, 1754, pp. 89, 96.
18   Ibid., p. 91.

desire for self-preservation deposited by God in his creation, and because the violation of natural laws leads to punishment, the optimisation of our happiness sets us into collaboration with the divine project. God instrumentalises instrumental reason. This reasoning formed the basis of Protestant apologetics for inoculation. Tissot writes:

When God created the universe, he established a certain number of physical laws that . . . are the best possible. When he created thinking beings, he engraved in them the foundations of moral laws. Since the first of these is that they should love themselves and seek their own happiness, he willed that . . . that which was to serve as the organ of this happiness should produce their well-being.[19]

In this ethic founded on natural law and concern for the self, medicine occupied an eminent place. Through its ability to define behaviour conducive to self-preservation, it rooted morality in physiology.[20] In the eighteenth century, many cultural anxieties were taken up by medical discourse: luxury, idleness, solitude, gluttony and, of course, masturbation. Medicine translated the moral disorders of the Enlightenment into physical ailments. The doctor saw himself as a more effective source of control than morality. In the *Encyclopédie*, Dr Menuret de Chambaud explained that 'to be a good moralist, you have to be an excellent doctor'.[21] With reference to onanism, Tissot explained that 'it is easier to turn away from vice by the fear of a present evil than by reasoning based on principles'.[22] Inoculation and masturbation belong to one and the same moment in the naturalisation of morality: 1718 saw the publication of *Onania*; 1721, the first inoculations in England; 1754, the start of the inoculation controversy in Paris; 1760, the publication of Tissot's *L'Onanisme*. Supporters of inoculation such as Mather, Tissot, Menuret de Chambaud and Bienville also wrote the main texts on the dangers of masturbation.[23]

19   Samuel Tissot, *L'Inoculation justifiée*, Lausanne: Bousquet, 1754, p. 107.

20   Thomas Laqueur, *Making Sex: Body and Gender from the Greeks to Freud*, Cambridge, MA: Harvard University Press, 1990; *Solitary Sex: A Cultural History of Masturbation*, New York: Zone Books, 2004.

21   Denis Diderot and Jean Le Rond d'Alembert (eds), *Encyclopédie, ou dictionnaire raisonné des sciences, des arts et des métiers*, Neuchâtel: Samuel Faulche, 1765, vol. 11, entry 'Œconomie animale', p. 360.

22   Samuel Tissot, *L'Onanisme ou dissertation physique sur les maux produits par la masturbation*, Lausanne: Chapuis, 1760, p. ix.

23   Cotton Mather, *The Pure Nazarite: Advice to a Young Man Concerning an Impiety and Impurity*, Boston: T. Fleet, 1723; M. D. T. de Bienville, *La Nymphomanie ou traité de la fureur utérine*, Amsterdam: Rey, 1771; Diderot and Le Rond d'Alembert (eds), *Encyclopédie*, vol. 10, article 'Manustupration', p. 51.

In both cases, medicine sought to guide behaviour by making the least risk a criterion of morality: the supposed dangers of onanism demonstrated the enormity of the sin, while inoculation was justified by the argument of the least risk.

By denaturalising instinctive moral disapproval, risk achieved a sort of moral coup, whose significance can hardly be overstated. According to the Boston physician William Douglass, the creation of an inner space based on probabilities and thus separate from collective norms risked dividing the religious and political community. The pastors' invocation of the natural order was a similar move to that of the Puritan zealots of the seventeenth century who claimed to be directly and personally inspired by God. Risk opened up the possibility of acting preventively, and in good conscience, on one's own body. It therefore seemed to be a prelude to other such initiatives, this time concerning the body politic.[24]

In Paris, inoculation and the new autonomy it allowed seemed to challenge the social order based on the absolute monarchy. A university doctor explained that, in authorising deliberate harm to the body, the principle of individual utility undermined 'the power that Princes have over the lives of men, as ministers of God': 'To suppose that the conduct of each individual in relation to his body is left to his own caprice . . . is repugnant to the idea of a well-ordered government.'[25]

2. Risk and the governance of autonomy

French propagandists for inoculation would indeed use probabilities to reorder authority within the body politic. At the Académie des sciences on 24 April 1754, the geometrician Charles-Marie de La Condamine read out a dissertation advocating inoculation. It was a huge success: all the periodicals reported on it, most of them dealing with the subject for the first time. Grimm commented: 'Mr de La Condamine has made a revolution.'[26] What was so revolutionary about the probabilistics of a medical technique?

First, risk made it possible to bypass the Faculty of Medicine, which (in theory more than in practice) had a monopoly on defining what counted

---

24   William Douglass, *Inoculation of the Small Pox as Practised in Boston*, Boston: Franklin, 1722, p. 12.

25   *Mémoire sur le fait de l'inoculation*, Paris: Butard, 1768, pp. 16–18.

26   *Correspondance littéraire, philosophique et critique de Grimm et de Diderot*, Paris: Furne, 1829, vol. 1, p. 454.

as legitimate remedies.[27] To the doctors who criticised his interference, La Condamine retorted that inoculation did not come under their remit: it was a 'complicated question that can only be resolved by measuring the greatest probability ... and we know that calculating risks is a matter for geometry ... the medical doctor is more capable of confusing than clarifying the question'![28] By posing inoculation as 'pure problem of probability', La Condamine also drew on his own expertise as a geometrician.[29] On his return from a long geodetic expedition to South America, he had earned a reputation in the public eye as a heroic scientist who had succeeded, in exotic and dangerous climes, in bringing about the triumph of European science's ideal of precision.

Second, risk marginalised the medical faculty by establishing the public itself as the proper judge of inoculation. In the 1750s, medicine was a province of the Republic of Letters. Medical debates belonged to the same textual universe as literature or politics: doctors, who had no specialised periodical until the founding of the *Journal de médecine* in 1755, published their articles in the generalist press. While the faculty did purport to impose regulations, it was the public sphere – the gazettes, posters and advertisements – that defined the commercial success of therapeutics.[30] Thanks to risk, the court of public opinion ceased to be a mere metaphor: by making it possible to assess the merits of a given medical procedure even while remaining ignorant of medicine, it established the public as the true judge of the question.

Finally, the use of probabilities had the merit of selecting the individuals worthy of sitting on this tribunal. As La Condamine saw things, reason was not the world's most widely shared attribute: it was necessary to exclude 'almost all women' and 'most men who have not had an education'.[31] This part of the population was radically delegitimised as a source of judgement: 'Their natural incapacity or lack of habit of exercising their reason, prevents them from stringing together two reasonings, and in their lives [they] have never arrived at a third consequence.' Furthermore:

---

27   Laurence Brockliss and Colin Jones, *The Medical World of Early Modern France*, Oxford: Clarendon Press, 1997.

28   Charles-Marie de La Condamine, *Histoire de l'inoculation de la petite vérole ou recueil de mémoires, textes, extraits et autres écrits sur la petite vérole artificielle*, Amsterdam: Société typographique, 1773, p. 492.

29   'Lettre de M. de La Condamine à M. Daniel Bernoulli', *Mercure de France*, April 1760.

30   Colin Jones, 'The great chain of buying: medical advertisement, the bourgeois public sphere, and the origins of the French Revolution', *American Historical Review*, vol. 101, no. 1, 1996, pp. 13–40.

31   La Condamine, *Histoire de l'inoculation*, p. 305.

'Out of a hundred women, of a hundred mothers, not one will be enlightened enough to see that she has to inoculate a beloved son.'[32] In a time
when women were increasingly characterised in terms of the dominance
of their feelings over their ability to reason, risk added to their exclusion
in an area – children's health – where their competence had nonetheless
been recognised. Risk thus created a narrow and sexist public sphere that
could be invoked as the supreme source of judgement.

The publication of La Condamine's essay marked the start of one of
the greatest controversies of eighteenth-century France.[33] From 1754
onwards, treatises, pamphlets and articles on inoculation followed in
rapid succession. Doctors, mathematicians, priests, moralists and philosophers all took sides, leading to a tangled debate and increasingly complex
arguments. For the defenders of inoculation and the philosophers of the
Enlightenment (and these two groups largely overlapped), this process
was an obstacle to their idea of good discussion. The *Encyclopédie* thus
distinguished between *controversy*, synonymous with (sterile and endless)
religious dispute and *discussion*. The latter 'expresses the action of *purifying* a subject of all those that may be foreign to it in order to present
it *clearly & uncluttered* of all those difficulties that *muddle it*'.[34] At issue
during the inoculation controversy was precisely the legitimacy of this
philosophy of disentanglement. The questions that would see casuists,
philosophers, doctors and mathematicians clash for a generation were
mixed: What was the moral and theological value of risk? Could medicine
go beyond cures? What does it mean to heal? What meaning should be
attributed to the desire to improve bodies? What would the consequences
be for the body politic?

Let us take a concrete example. Inoculation was often criticised for
'doing an evil to obtain a good' or 'inserting an ill into a healthy body'. This
ethical problem led to theoretical discussions on the nature of smallpox:
could its generalised presence be explained by the existence of an innate
germ, or was it a purely contagious disease like syphilis? The innate
germ doctrine, which held smallpox to be a crisis inherent to human

---

32   Charles-Marie de La Condamine, 'Lettre de M. de La Condamine, à M. l'abbé
Trublet', *Année littéraire*, 6, 1755, p. 17.
33   This period saw a politicisation of the public sphere. Through a series of crises
(the denial of the sacraments to the Jansenists, Damiens's assassination attempt on Louis
XV, the censorship of the *Encyclopédie*), public opinion became, in the political culture
of the Ancien Régime, the authority by which the sovereign was judged. Roger Chartier,
Les Origines culturelles de la Révolution française, Paris: Seuil, 1990.
34   Diderot and Le Rond d'Alembert (eds), *Encyclopédie*, vol. 4, p. 1034, entry
'Discussion'.

nature, in the manner of puberty, had been abandoned by physicians by the 1750s. There was now a consensus around the theory of contagion. However, with the inoculation controversy, medical opinion seemed to be taking a step backwards.[35] The innate germ theory allowed moral objections to disappear: on this reading, inoculation did not introduce a disease but destroyed a vice already present in the body; if a person died from inoculation, it was possible to blame a particularly virulent germ to which the inoculated person would have succumbed sooner or later in any case. The innate germ theory attributed accidents to the body of the inoculated person rather than to the inoculator's lancet. For many propagandists, it seemed simpler to resort to this clumsy theory than to get bogged down prattling on about ethics.[36] But because it seemed to *predestine* death or survival, the innate germ theory also had troubling theological connotation in France. In the Faculty of Medicine, which was deeply divided on the Jansenist question, propagandists and opponents of inoculation accused each other of making unorthodox statements about predestination.[37]

Inoculation, like a Möbius loop, gave the illusion of having two sides, one medical and the other ethical, which, on examination, turned out to form a continuous surface of problems and arguments. In the words of the time, it was 'impossible to separate the moral from the physical'. In June 1763, the Paris *parlement* (a major jurisdiction in France) consulted the Faculty of Medicine on 'the physical side' of the question alone, reserving the moral issue for the theologians at the Sorbonne.[38] But, in 1768, as the debate stalled, an anti-inoculation doctor proposed a change of strategy: since the faculty could not reach any agreement on the medical value of this innovation, it should judge it according to the moral laws of medicine. These laws were based on the distinction between therapeutics and hygiene. The former, whose object is the sick body, may use 'unnatural means' (purging, bleeding, amputations) because it takes the body 'in a

---

35   Tissot proposed a simple analogy: milk that curdles with a drop of acid. The potential for curdling is internal to the milk, but it is only expressed thanks to a germ that comes from the outside. Inoculation was a means of exhausting the potential for disease: just as milk that has already curdled no longer curdles, the person who had been inoculated would not get smallpox. See Tissot, *L'Inoculation justifiée*, p. 33; Charles Rosenberg, *Explaining Epidemics and Other Studies in the History of Medicine*, Cambridge: Cambridge University Press, 1992.

36   M. D. T. de Bienville, *Le Pour et le contre de l'inoculation ou dissertation sur les opinions des savants et du peuple sur la nature et les effets de ce remède*, Rotterdam, 1768.

37   Guillaume J. de l'Épine, *Rapport sur le fait de l'inoculation*, Paris: Quillau, 1765, pp. 23–7.

38   'Arrêt de la cour de parlement sur le fait de l'inoculation', 8 June 1763.

state of necessity'. The second is aimed at the healthy body and seeks to maintain it in this condition by using 'natural means': rest, diet, environment, and so on. Inoculation has no place in this dualism because it is an unnatural means used on a healthy body. To authorise inoculation would be to endorse the existence of a third type of medicine, one based on the 'transformability' or 'mutability' of the human body. Doctors would be led down a dangerous interventionist path aimed at endlessly improving the human body at the behest of their patients.[39]

This medical casuistry extended transcendence to the body: the believer was no longer simply accountable for his life, which he had to manage like a good merchant calculating his chances, but for the very physicality of his body. This body was entrusted to him in its existing condition, and he could not alter it at will. Taking up the Pauline vision of a body that does not belong to the person who inhabits it – 'Do you not know that your bodies are temples of the Holy Spirit, who is in you, whom you have received from God? You are not your own' – the anti-inoculator insisted that our body 'has not been given to us to serve our experiences'.[40] In the religiosity of the Counter-Reformation, centred on the suffering body and the virtues of pain, it was unacceptable to use morally dubious medical practices to escape illness. Was illness not itself a worthwhile penance? Moreover, the embodiment of God in Christ justified the theological appreciation of the healthy body as an absolutely perfect structure. Any wilful alteration of it seemed sacrilegious.[41]

The purpose of risk was precisely to purify the debate, to free inoculation from the indeterminacy of moral and medical disputes in order to present it 'clearly & uncluttered' to the reasonable public and its judgements. According to La Condamine, inoculation 'is no longer a question of morality or theology, but a matter of calculation: let us take care not to make a matter of conscience out of a question of arithmetic'.[42] The geometrician responds to moral misgivings with theorems – 'Should a father voluntarily expose his son? Yes, and I can prove it' – and concluded his reasoning with: 'It is therefore demonstrated to the fullest extent of the term.' QED. By freeing the individual from moral and theological constraints and leaving calculating reason as the sole determinant of behaviour, risk

---

39   *Mémoire sur le fait de l'inoculation*, p. 46.
40   Corinthians, 6, 19; *Mémoire sur le fait de l'inoculation*, p. 27.
41   Jacques Gélis, 'Le corps, l'église et le sacré', in Alain Corbin, Jean-Jacques Courtine and Georges Vigarello (eds), *Histoire du corps*, Paris: Seuil, 2005, vol. 1, pp. 17–107.
42   Charles-Marie de La Condamine, 'Mémoire sur l'inoculation de la petite vérole', in *Histoire de l'Académie royale des sciences*, Paris: Imprimerie royale, 1754, 1759, pp. 649–55.

was meant to establish a single admissible way of looking at inoculation. This had radical political consequences: in a society of rational beings, there can be only one way of behaving. Risk was supposed to produce a society of free and rational individuals: free to choose risk, and indeed obliged to take risk because they are rational. It also provided for both the autonomy of human conduct from moral standards, and an internal discipline that would keep this behaviour under control.

3. From the viral contract to the social contract

The creation of calculating subjects was at the heart of a political project. It seemed to be an essential condition for the functioning of a society made up of sovereign individuals: unlike moral and religious principles, which are irreducibly plural and contradictory, and unlike virtue, pity or benevolence, the ability to calculate seemed to be the only faculty sufficiently shared to form a broad political community and a consensus on laws. Once the transcendental foundations of the social order based on absolute monarchy had been revoked, it appeared that autonomous individuals could be governable, provided that they became calculating subjects.[43]

It may be that inoculation and the new attention paid to the concept of risk played an important role in the development of utilitarian philosophy. The greatest happiness of the greatest number was the inoculators' maxim long before Jeremy Bentham made it the touchstone of all legislation. In 1754, Tissot justified the losses caused by inoculation as follows: 'Everything must be calculated in our conduct . . . Those to whom the fate of men is entrusted must always reduce their calculation to the common sum.'[44] According to Claude Adrien Helvétius, another champion of inoculation, any law had to meet the requirement of 'public utility, i.e. that of the greatest number'. At the height of the inoculation controversy, several key texts appeared: Helvétius's De l'Esprit (1758), Rousseau's Du Contrat social (1762) and Cesare Beccaria's treatise On Crimes and Punishments (1764). This was no mere coincidence: the controversy specifically concerned the question at the heart of the utilitarian philosophy in the

---

43   Élie Halévy, La Formation du radicalisme philosophique, Paris: Alcan, 1901, vol. 1; Mary Poovey, A History of the Modern Fact: Problems of Knowledge in the Sciences of Wealth and Society, Chicago: University of Chicago Press, 1998, pp. 144–50.

44   Tissot, L'Inoculation justifiée, pp. 94–5 (citing Charles Duclos, Considérations sur les mœurs de ce siècle, Paris: Prault, 1751).

making, namely the possibility of making the individual calculation of risks and benefits into the basis of public morality.[45]

In *On Crimes and Punishments*, Beccaria took as his starting point a potential criminal who considers the *risks* of his misdeed. For a penalty to be effective, the risk need be only slightly greater than the benefit derived from the crime. At the time Beccaria developed this penal theory, he was contributing to the Milanese journal *Il Caffè*, which had published a long article on inoculation that presented the relationship between risk and hope as the motive for decisions.[46] The type of power that was invented during the inoculation controversy, and which Beccaria was one of the first to reflect upon, set its sights not on behaviour as such, but rather on the motives for behaviour: the sovereign must act indirectly on the motives for decisions, through education and the law, so as to produce calculating subjects and direct their individual calculations towards the maximal public good. Similarly, according to Baron d'Holbach, also a proponent of inoculation, the good sovereign should be content to enlighten the citizens, and to 'bring them by gentleness to the reason they ignore'. This soft-power approach, effectively allowing individuals to govern themselves, depended on the formation of probabilistic and sensitive subjects, educated to perceive small risks and to act accordingly.

Inoculation was emblematic of this gentle, indirect way of governing. Until the end of the eighteenth century, it seemed politically and morally impossible to impose it on individuals. In 1764, at the Paris Faculty of Medicine, Dr Antoine Petit concluded his report in favour of inoculation with a declaration of impotence: 'Life and health are among the things ... that a wise government abandons to the discretion of individuals ... The interest in one's own preservation is the strongest of all laws, from which it follows that the public must be left to make the law for itself.'[47] As inoculation seemed both essential to the public good and to lie outside the sphere of government action, it had to be generalised not through the enactment of positive norms, but through persuasion. In 1802, Bentham used the example of the story of inoculation to demonstrate the effectiveness of gentle government: 'Should inoculation have been established by direct law? No, undoubtedly ... it would have brought fear to a multitude

---

45   Halévy, *La Formation du radicalisme philosophique*.

46   Pietro Verri, 'Sull'innesto del vajuolo', *Il Caffè: o sia, brevi e vari discorsi già distributi in fogli*, vol. 2, 1764, pp. 452–523.

47   Antoine Petit, *Premier Rapport en faveur de l'inoculation, lu dans l'assem- blée de la faculté de médecine de Paris en l'année 1764, et imprimé par son ordre*, Paris: Dessain, 1766, p. 144.

of families. The practice became universal in England through public discussion of its advantages.'[48] The gentle and indirect liberal government of bodies relied on individuals' inner convictions and thus on the 'judgement of the public'; its effectiveness depended on its ability to put public debate at the service of its projects.

The notion of risk also allowed Rousseau to resolve the problem that the death penalty posed for the theory of the social contract: 'It may be asked how individuals who have no right to dispose of their own lives can transmit to the sovereign this right which they do not possess.'[49] In a manuscript from 1756, Rousseau reflected on the difference between peril, danger and risk. Risk refers to 'a danger to which one exposes oneself *voluntarily* and with some hope of escaping in order to obtain something that tempts us more than the danger frightens us'.[50] The solution to the paradox of capital punishment lies in this lexical analysis: it is true that the individual does not 'dispose of his own life' (suicide is forbidden), but 'every man has the right to *risk* his own life in order to preserve it'. The individual can therefore subscribe to a social pact that maximises his security, even if the sovereign has the right to kill him. A point of capital importance is that, before signing the pact, the individual does not know whether he will be an honest person or a murderer: 'It is in order not to be the victim of an assassin that a man consents to die if he becomes one.'[51] The *Social Contract* was written at the same time as the debate on the legitimacy of inoculation. However, to justify the risk of death, the promoters of inoculation also used the idea of equality in misfortune and ignorance of outcomes. A doctor in Strasbourg explained that inoculation was not immoral because everyone freely accepted the risk and because no one knew who would die: 'One does not choose one man out of 1,000 to plunge a knife into his blood, one does not determine who will die; each man can tell himself that his risk of dying is 1 in 1,000.'[52] Like the death penalty in the social contract, inoculation was legitimate because each of the contracting parties shared equally in the gains and losses.

---

48   Jeremy Bentham, *Traités de législation civile et pénale*, Paris: Bossange, 1802, vol. 3, p. 360.

49   Jean-Jacques Rousseau, *The Social Contract and the First and Second Discourses*, New Haven, CT: Yale University Press, 2002, p. 176.

50   Bruno Bernardi, 'Le droit de vie et de mort selon Rousseau: une question mal posée?', *Revue de métaphysique et de morale*, 2003, pp. 89–106, citing ms R. 16 in the Neuchâtel university library.

51   Rousseau, *The Social Contract*, p. 177.

52   François-Antoine Hertzog, *Réfutation de la réfutation de l'inoculation*, Strasbourg: Christmann et Leurault, 1768, p. 108.

4. State rationality and the practice of the slave master

A second way of looking at inoculation was not from the viewpoint of the candidate for the operation, but that of the sovereign seeking to increase his number of subjects. In the context of the Seven Years' War and the ongoing French–British rivalry, the argument in terms of population numbers made it possible to vest inoculation with the national interest: according to Antoine Petit, the leader of the partisans of inoculation at the Faculty of Medicine, any European nation that refused inoculation would inevitably be subjugated within the course of a few centuries.[53]

In 1760, the Basel mathematician Daniel Bernoulli developed a mathematical model simulating the effect on the population of systematically inoculating newborn babies. The aim was to show that the sovereign could optimise his population, increase its numbers and change its generational structure. The question of risk became marginal here: if all children were inoculated at birth, the population of newborns would immediately be reduced (risk of 1 in 200 deaths) but 'this deduction is very small and could almost be overlooked for the sake of the totality, which alone deserves the prince's attention'. The main advantage of inoculation was that the human cost of immunisation was offloaded onto infants: 'The loss falls only on those children who are useless to society, and all the gain is passed on to the fertile age group.'[54] At the age of twenty-five (an age when individuals were of use to the state, which needed soldiers, workers and progenitors), inoculation resulted in a gain of 79.3 individuals out of a cohort of 1,300 children. In an explicitly agricultural metaphor, the sovereign had to manage his population like a good agronomist takes care of his crop yields: he must pay the utmost attention to 'the age of the harvest'.[55]

This essay was emblematic of a new political rationality that, following Michel Foucault, has come to be known as *biopower*. In the eighteenth century, with the emergence of demography, the subjects of a kingdom were rethought as forming a *population*: a statistical entity that could be counted, calculated and forecast. Philosophers, economists and doctors called for the active management of this new object: the laws governing

53   Petit, *Premier rapport en faveur de l'inoculation*, p. 81.
54   Daniel Bernoulli, 'Nouvelle analyse de la mortalité causée par la petite vérole et des avantages de l'inoculation pour la prévenir', *Mémoires de mathématique et de physique tirés des registres de l'Académie royale des sciences*, 1760, p. 34.
55   Daniel Bernoulli, 'Réflexions sur l'inoculation', *Mercure de France*, June 1760, p. 178.

it (birth rate, mortality, life expectancy) had to be studied, and the levers for influencing its development identified (political economy, taxes, inheritance laws, health and environmental conditions), all with the aim of maximising the population in the name of the realm's wealth and power. The object of power shifted: it was no longer a set of subjects that needed to be 'disciplined and punished', but a population that, subject to natural laws that the sovereign could not change, needed to be governed indirectly, mainly through political economy. The liberal management of hunger was an example of this new type of power. According to the physiocratic economists of the 1760s, in the event of poor harvests, the sovereign had to accept the rise in the price of grain and the risk of food shortages in order to give free rein to the expectations of farmers, who would increase their production. By allowing a shortage to develop, the sovereign would reduce the risk of famine in general.[56] The prince's rationality in Bernoulli's essay was similar: rather than containing small-pox through quarantines, he should accept inoculation and the risk of individual accidents in order to control the epidemic phenomenon and reduce its impact on the population as a whole.

This probabilistic population management did not have applications in public health policies in the eighteenth century. However, it was at the heart of the capitalist management of life in two areas: life annuities and slave populations.

Let us begin with life annuities. Generally issued by governments, they provided a regular income in exchange for capital. Their high yield (around 10 per cent) assumed amortisation (the death of the annuitant) at twenty years, slightly less than life expectancy at birth in eighteenth-century France. However, from 1770 onwards, Geneva bankers began to optimise the income from the life annuities issued *en masse* by the French state. They selected children in robust health, usually girls aged seven or eight, and bought annuities in their names for several million livres. To help to raise funds, they consolidated these annuities to sell their customers a less risky financial product, based not on the life of an individual but on the lives of a group of thirty young girls, known as the 'thirty immortals of Geneva'. The annuity thus lost its personal character and could be resold in all the European markets, particularly in Holland.

The lives of the 'thirty immortals' were entirely subordinated to the imperative of longevity. To avoid the risks of childbirth, they were required

56   Michel Foucault, *Security, Territory, Population, Lectures at the Collège de France, 1977–1978*, New York: Palgrave Macmillan, 2007.

to remain celibate. The Geneva physician Théodore Tronchin, Europe's most prestigious inoculator, was responsible for selecting them, maintaining their health and inoculating them. Thanks to these precautions, of the first 'thirty immortals' selected in 1772, twenty-eight were still alive twenty years later. This combination of financial and medical innovation contributed directly to the financial collapse of the French monarchy: in 1789, out of 530 million livres of revenue, 162 million went to Geneva to honour the annuities![57]

Although no sovereign followed Bernoulli's project, his essay did have a concrete model: that of the slave master. Inoculation had been commonly practised in North Africa, on the coasts of Guinea and in Sudan, since at least the seventeenth century; slave traders soon borrowed this indigenous technology in order to protect their commodity.[58] As early as 1723, the Royal African Company hired an inoculator at its slave trading posts.[59] Inoculation was profitable in three ways: it increased their market value; reduced losses during the crossing; and avoided slaves being forced into quarantines on the sugar islands. Inoculation made the slave trade more fluid, reduced its risks and increased capital yields.

In the colonies, inoculation was practised on an altogether different scale than in metropolitan France: in 1756, while there were only a few inoculated people in Paris, the governor of the Îles de France et de Bourbon (Mauritius and Réunion) had 400 Blacks inoculated.[60] In the plantations of Saint-Domingue, the first inoculations were carried out in 1745, and the practice became widespread in the 1770s.[61] In Louisiana, during the epidemic of 1772, more than 3,000 slaves were inoculated in just a few months.[62]

---

57   Herbert Luthy, *La Banque protestante de la révocation de l'édit de Nantes à la Révolution*, Paris: SEVPEN, 1961, vol. 2, pp. 465–500.

58   See Eugenia W. Herbert, 'Smallpox inoculation in Africa', *Journal of African History*, vol. 16, no. 4, 1975, pp. 539–59. According to Cotton Mather, slaves in Boston were generally inoculated. See George Kittredge, 'Lost works of Cotton Mather', *Proceedings of the Massachusetts Historical Society*, 45, 1912, p. 431.

59   Larry Stewart, 'The edge of utility: slaves and smallpox in the early eighteenth century', *Medical History*, 29, 1985, pp. 54–70. According to Herbert S. Klein (*The Middle Passage*, Princeton, NJ: Princeton University Press, 1978), inoculation partly explains the reduction in slave mortality during the Atlantic crossings from the 1750s onwards.

60   Jean-François Bougourd, 'Histoire de l'inoculation à Saint-Malo', *Journal de médecine*, 34, 1770, pp. 134–51.

61   Gabrien Debien, *Les Esclaves aux Antilles Françaises, XVIIe–XVIIIe siècles*, Fort-de-France: Société d'histoire de la Martinique, 1974, pp. 313–16.

62   M. Le Beau, 'Observations sur quelques inoculations faites à la nouvelle Orléans, dans la Louisiane', *Journal de médecine*, 40, 1773, pp. 501–4.

Probabilistic rationality could be applied to slave populations without any constraint. In the West Indies, the doctor Jean-Baptiste Leblond inoculated hundreds of slaves by proposing an insurance scheme to their masters: the operation was billed 'at the rate of twenty francs per head' and the doctor undertook to pay 1,000 francs in the event of death.[63] The inoculation of slaves remained quite absent from the French debate: it caused no scandal, but also offered no proofs. Because Black bodies were considered too different to be able to draw any medical conclusions from them, the propagandists of inoculation made little reference to this practice, except to give it as an example of a calculating rationality that the French public would do well to follow.[64]

5. The moral failure of risk

The contrasting fortunes of inoculation on the plantations and in mainland France show how far the concept of risk failed to convince autonomous subjects. In 1758, after four years of propaganda, La Condamine counted less than 100 inoculated people in Paris; ten years later, there were just over 1,000 in the whole of France. There were many reasons for this failure. They were partly moral. The assertive tone of the defenders of inoculation and Voltaire's sarcastic remarks on the *parlement*'s ruling of June 1763 requesting the opinion of the Faculty of Theology might lead us to believe that ethical critiques were marginal or outdated. Yet this was not the case: all the French judicial institutions that were confronted with the problem of inoculation declared it an illicit practice.

In 1754, a doctor from Nancy asked the public prosecutor for authorisation to inoculate. This latter was greatly embarrassed: who would be held responsible in the event of an accident? The child's parents? The doctor who prepared the patient? The surgeon who made the incision? Puzzled, the public prosecutor sought the counsel of the Advocate General of the *parlement* in Paris, Joseph Omer Joly de Fleury. The latter's response was categorical: 'As this practice is dangerous in itself, all those who dare to carry it out should be severely prosecuted.'[65] Before issuing his ruling in

---

63   Jean-Baptiste Leblond, *Voyage aux Antilles et à l'Amérique méridionale, commencé en 1767 et fini en 1802*, Paris: Arthus-Bertrand, 1813, p. 355.

64   Londa Schiebinger, 'Human experimentation in the eighteenth century: natural boundaries and valid testing', in Lorraine Daston and Fernando Vidal (eds), *The Moral Authority of Nature*, Chicago: University of Chicago Press, 2004, pp. 385–409.

65   BNF, ms Joly de Fleury, ms 307, fol. 164. Letter from Joseph Omer Joly de Fleury to the procureur de Nancy, 28 December 1754.

June 1763, Joly de Fleury had consulted the Abbé Gervaise, at the time the Syndic of the Sorbonne, to ascertain the theologians' position. His reply left no doubt as to the procedure's final outcome: 'If experience shows that even though inoculation is generally successful, it nevertheless has its dangers ... then we would be obliged to decide that this practice is not permitted.'[66] By 1763, the opponents of inoculation seemed to have won the moral debate: the probabilistic justification of deliberately running the danger of death was bluntly rejected by the jurists and casuists at the Sorbonne.

In fact, describing inoculation as a chosen, uniform, individual and therefore moral risk oversimplified the nature of the danger. The moral problem lay not so much in choice as in the lack of it: that is, the lack of choice by the children inoculated at an early age to avoid wasting the cost of an education, and the lack of choice by those around them, forced to breathe infected air. Inoculation redistributed the risks unequally: by having themselves inoculated, the rich escaped the common epidemic fate and exposed the common people to increased contagion. Worse still, by increasing contagion, inoculation placed constraints on future generations: 'If you allow inoculation, you will make the smallpox eternal and universal. The more you inoculate, the more you'll want to be inoculated.'[67] The dilemma posed by inoculation is similar to the prisoner's dilemma: through the sum of individual choices, the community would be driven to choose the worst solution. It was precisely the irreversible nature of inoculation that made the quarantine strategy attractive. Jean-Jacques Paulet, a doctor at the University of Montpellier, emphasised that inoculation represented a leap into the unknown: by multiplying the sources of infection, it 'would place men in a desolation that it will no longer be possible to remedy'.[68]

The inoculation of Louis XVI in 1774 is presented by historians as a symbol of the inoculists' victory in France. However, at the same time, numerous ordinances banned inoculation from urban areas in the name of public health. For instance, an important judicial court in Lyon was careful to distinguish between royal inoculation, a symbol of bravery, and the public good, which demanded the banning of this practice.[69] From

---

66   Ibid., ms 577, fol. 184.

67   Pierre-Louis Le Hoc, *L'Inoculation de la petite vérole renvoyée à Londres*, The Hague, 1764, p. 102.

68   Jean-Jacques Paulet, *Avis au public sur son plus grand intérêt ou l'art de préserver de la petite vérole*, Paris: Ganeau, 1769, p. iii.

69   BNF, ms Joly de Fleury, ms 2420, fol. 280, Jugement de la sénéchaussée de Lyon, 19 May 1778.

the 1770s onwards, Paulet's quarantine projects had the wind in their sails. In 1768, his *Histoire de la petite vérole* was a great success: Voltaire, who had been the first philosopher to encourage inoculation, claimed that he had changed his mind after reading this work.[70] Paulet demonstrated the potential success of quarantines: by taking strict precautions, remote villages had been able to protect themselves from epidemics. Some localities around Montpellier had not experienced smallpox for over thirty years. Inoculation and quarantine seemed to correspond to different political forms: individualistic, poorly governed countries such as England had to resort to inoculation; conversely, quarantines were desirable in well-administered kingdoms such as France. *Parlements* banned inoculation in cities: Paris in 1763; Metz in 1765; Saint-Omer in 1776; and Dijon, Lyon and Rennes in 1778.

Inoculation retained its reputation as a selfish practice. Kant offered a telling summary of this moral dilemma. In 1800, a young aristocrat asked him whether inoculation was legitimate. In his reply, the philosopher distinguished between two laws: the subjective law of pragmatic action and the moral maxim, the law that determines the probable advantages and the law that prohibits doing harm in the hope of obtaining some benefit.[71] And the philosopher of audacity, the one who had defined the Enlightenment as humanity's coming of age, finally recoiled faced with the bodily consequences of autonomous reason. For the Enlightenment, inoculation posed a problem because it seemed to upend its values and transform them into evils. The desire to act on a healthy body, to transform it under the pretext of utility, illustrated the dangers of an individualistic ethic prone to excess in its conquest of autonomy.

6. Risks and sensibility

The concept of risk also failed at a psychological level: it seemed to describe too meagre a vision of reason, far removed from both the ideal of sensibility and the aristocratic ethos of the bold feat. The fact that the first to be inoculated were almost all young nobles testified to the failure of the probabilistic argument: the danger of inoculation was not interpreted as a small risk that called for a rational decision, but rather as an

---

70   'Lettre de M. de Voltaire à Paulet au sujet de l'histoire de la petite vérole', *Mercure de France*, July 1768, p. 149.

71   Immanuel Kant, 'Réflexions sur l'inoculation', *Écrits sur le corps et l'esprit*, Paris: Flammarion, 2007.

aristocratic exploit. The allegories for inoculation bore witness to this: a winged 'Victory' was engraved on a stele in honour of the first woman to be inoculated in Denmark, and it was in the shadow of this stele that the Prince of Denmark in turn received the inoculation.

The Duchess of Amon received a medal bearing the laurels of courage for having had her three children inoculated simultaneously (*tres liberi simul inoculati*), in what was deemed a great stroke of audacity.

In Paris in 1756, Tronchin's inoculation of the Duc d'Orléans's children was interpreted as a heroic deed. Louis Poinsinnet composed a *Poème à Monseigneur le Duc d'Orléans* in epic style, praising the courage of his protector.[72] When the inoculated reappeared at the Opéra, they were

---

72   Louis Poinsinet, *L'Inoculation. Poème à Monseigneur le Duc d'Orléans*, Paris, 1756.

applauded by the audience. When the Duc de Gisors asked his father for permission to be inoculated, his father refused on the grounds that it was a voluntary danger 'in which there is neither honour nor glory to be attained'.[73]

According to the Protestant casuistry of inoculation, reason derived from the laws of nature should govern human sentiments. In Paris in the 1760s, this injunction clashed with the new emphasis on sensibility: novels and aesthetic and moral treatises instead invited readers to introspection, to the exploration of their inner selves and to the discovery of unknown emotional capacities. The morality of feeling undermined the casuistic justification based on the natural order and instead reactivated the Augustinian moral figure of the inner light. The Earl of Shaftesbury and Francis Hutcheson thus criticised Lockean morality and instead postulated the existence of an 'inner eye': of innate feelings of benevolence, pity and love that enable us to perceive the moral good. The theme of sensibility was a special hindrance to the inoculists' efforts as it primarily concerned the family and the education of children. In the mid-eighteenth century, the way in which filial relationships were evoked changed in character: as theories of education multiplied, feelings of tenderness and love became the legitimate way to describe these relationships; they were what gave value to being a mother or a father. The family was increasingly conceived as a sphere of intimacy based on feelings that ought to be isolated from the relations of calculation that governed the economic sphere.[74]

The opponents of inoculation seized on sentimental discourse in order to reject the opinion of high society and urge the people to pass judgement for itself. The author of *Désaveu de la nature* noted 'a striking difference ... the little people, or what we call good folks, have only one feeling, and it is against. The mighty, or those who set the tone and those who receive it, have only one voice, in favour of inoculation.'[75] Tears also played an important role in communications opposed to the practice. The *Désaveu de la nature* was a tearful poem recounting the despair of parents overwhelmed by the death of their son from inoculation. A few years later, an *académicien* from Lyon published an elegy for his daughter, who

---

73   *Mémoires du Duc de Luynes sur la cour de Louis XV*, Paris: Firmin-Didot, 1864, vol. 15, p. 21.

74   Daniel Mornet, *La Pensée française au XVIIIe siècle*, Paris: Armand Colin, 1926; Philippe Ariès, *L'Enfant et la vie familiale sous l'Ancien Régime*, Paris: Seuil, 1975.

75   Gabriel de Saint-Aubin, *Le Désaveu de la nature. Nouvelles lettres en vers*, Paris: Fetil, 1770, pp. 2–3. On the democratic function of feeling: William Reddy, *The Navigation of Feeling: A Framework for the History of Emotions*, Cambridge: Cambridge University Press, 2001.

had died after the operation. According to the *Correspondance littéraire*, it 'brought tears to the eyes of the whole of France'.[76] In the morality of feeling, the loss of a child through inoculation took on the status of a catastrophe that made the parents lose all meaning in life.

The fear of remorse was undoubtedly the most persistent obstacle to inoculation. Knowing that the odds were in her favour, Madame Rolland nevertheless hesitated to have her daughter inoculated, as she wanted to minimise her potential regrets: 'I would easily decide for people indifferent to me, as there are many probabilities in favour; but I would blame myself all my life for having exposed my child to exceptions to this good and I would rather that nature killed her than for this to be done by me.'[77] The psychology of high society was proving decidedly more complex than the 'weighing of risks' which La Condamine called for.

D'Alembert offered an in-depth analysis of the psychological failure of risk. Instead of calculating the objective risks of inoculation and natural smallpox, as La Condamine and Bernoulli had done, he tried to give mathematical form to the procrastination involved. In his view, the aim of probability theory is not to rectify sensitivities but to try to describe them rigorously – an infinitely more delicate task.[78]

First, it was necessary to take into account the preference for the present. But there is no theorem that tells us how to compare an immediate risk (inoculation) with the sum of the risks of dying of natural smallpox at each age, risks that 'weaken with distance, by the distance at which they are seen, a distance that makes them uncertain and softens the view of them'.[79]

Second, it makes no sense to compare risks in isolation, because:

If one person has to run risks A, B, C, D, E and another has to run risks a, B, C, D, E, it is obvious that to compare the total risk run by these two individuals, we have to compare the sum of risks $A + B + C + D + E$ with $a + B + C + D + E$, and not just risk A with risk a . . . The advantage

76 François Métra, *Correspondance secrète, politique et littéraire*, London: J. Adamson, 1787, vol. 14, 28 May 1783.

77 Manon Roland, *Lettres de Mme Roland*, Paris: Imprimerie nationale, 1902, vol. 2, Mme Roland to M. du Bosc, 6 April 1788.

78 Lorraine Daston, *Classical Probability in the Enlightenment*, Princeton, NJ: Princeton University Press, 1988, pp. 82–9; Hervé Le Bras, *Naissance de la mortalité, l'origine politique de la statistique et de la démographie*, Paris: EHESS, 2000, p. 330.

79 Jean Le Rond d'Alembert, 'Réflexions philosophiques et mathématiques sur l'application du calcul des probabilités à l'inoculation de la petite vérole', *Mélanges de littérature, d'histoire et de philosophie*, Amsterdam: Chatelain, 1767, vol. 5, p. 320.

is further diminished by this consideration, since the ratio of a + B + C + D + E to A + B + C + D + E is closer to equality than that of a to A.[80]

This would mean, for example, that for an individual who was already taking a lot of risks (a sailor or a miner, for example), there would be a lesser appreciation of the benefits of inoculation.

D'Alembert also drew on an argument dear to the opponents of inoculation, concerning the difference in the nature of the risks: losing one's son to natural smallpox did not have the same moral consequences as losing him to inoculation. A manifestly immoral probabilistic argument is false from the viewpoint of subjective probabilities.

Finally, d'Alembert disentangled the state's interest from that of the individual, and emptied Bernoulli's populationist arguments of their normative substance. Bernoulli had calculated that inoculation would increase average life expectancy by four years. D'Alembert showed that this is not enough to force an individual decision. Consider an imaginary world where the longest lifespan is a hundred. Smallpox is the only fatal disease; every year, it kills an equal number of men. The average lifespan would therefore be fifty. Inoculation saves men from smallpox, so the inoculated are sure to live to the age of a hundred – but the inoculation itself is fatal one time in five. If all children are inoculated 'while suckling', then the average life expectancy becomes eighty years. Although the state gains (the sum of life is increased), 'there would perhaps not be a citizen sufficiently brave or reckless to expose himself to an operation in which there would be one chance against four of losing his life'.[81] The populationist argument only applies to 'the state that considers all citizens indifferently', but it is a 'political chimera' because it 'cannot force any citizen to decide to adopt it'.

D'Alembert demonstrates perfectly why the concept of risk is unable to produce any generally shared conviction. Since individuals are not uniform, risk perception cannot be straightforward: 'The appreciation will be very different for each individual, in relation to his age, his situation, his way of thinking and feeling, the need that his family, his friends, his fellow citizens may have for him . . . there will perhaps not be two individuals who appreciate it in the same way.'[82] Far from standardising judgements, risk sets different men's consciences at odds with one another.

---

80   Jean Le Rond d'Alembert, *Opuscules mathématiques*, Paris: Briasson, 1768, vol. 4, p. 336.

81   D'Alembert, *Opuscules mathématiques*, 1761, vol. 2, pp. 35–8.

82   D'Alembert, 'Réflexions philosophiques', p. 325.

7. Risk in the salons

Finally, the notion of risk failed to convince people for practical reasons. First, the inoculation revealed the difficulty of harnessing aristocratic bodies to probabilistic rationality. Certain reactions to the inoculation of the young Louis XVI underlined the gap between the aristocratic conception of the body and the seriality implied by the concept of risk. Applying the calculus of probabilities to the king appeared scandalous, because it meant setting him on the same level of accounting as the other individuals who were inoculated.[83]

Second, the lists of inoculations from which risk was calculated did not inspire confidence. The inoculators were suspected of expunging them by attributing their failures to some widespread epidemic or to a hidden defect in the subject. The probabilistic argument was misleading: as the inoculators only accepted individuals in good health, it amounted to comparing the mortality from common smallpox with that of a selected elite. Worse still, by taking on healthy subjects, the inoculators systematically increased the mortality rate for natural smallpox, since it 'fell on the cursed'.[84] In any case, by trumpeting their successes and printing and distributing lists of operations, they were behaving like the 'charlatans' (therapists who had not taken their degrees) who displayed the tables of their 'treatments' around the Pont-Neuf. One opponent of inoculation quipped: 'It is only the empirics and inoculators who give lists of their patients who recover.'[85]

Third, the assessment of inoculation was not just a matter of life and death; it was also played out on the surface of bodies, according to the discomforts and scars that were expected. Smallpox had many faces. Some so-called discreet forms were benign. Similar to chickenpox, they left no scars. At the other end of the spectrum, confluent, putrid or hae-morrhagic forms were usually fatal. In between, a continuum of varied cases, symptoms and scars lent itself poorly to statistical treatment. The problem, however, was to determine where the inoculated smallpox fell

---

83 For example, the short play *L'Ombre de Vadé ou le triomphe de l'inoculation*, Paris, 1774, depicts a quarrel between two fishmongers at Les Halles. Against Franchon's probabilistic arguments in favour of inoculating Louis XVI, Margot retorts: 'Splendid, then / Because if only one in a million would die of it, / That would be enough to defend / Our great prince from undertaking / This operation, then, / Because our King alone is worth more than a million, / Since he makes the whole of France happy.'

84 De L'Épine, *Rapport sur le fait de l'inoculation*, p. 65.

85 *Lettre apologétique de M. Gaullard fils, pour servir de réponse à la lettre de M. de la C. insérée dans le Mercure du mois de mars 1760*, Paris, 1760, p. 37.

on this spectrum. Ideally, the best way to find out was by examining the faces of inoculated people or, failing that, by listening to and reading descriptions of inoculated bodies.

Finally, the risk lacked specificity. Accepting its logic required thinking about the body as generic, despite differences in age, temperament, the 'humours' and medical history. The social elite preferred to make their own judgements using their networks of acquaintances, rather than relying on lists whose apparent objectivity concealed subjective, self-interested medical opinions. The role of social connections in producing trust is clearly seen in the web-like spread of inoculation. For example, the Pisan inoculator Angelo Gatti frequented the salon d'Holbach, where he found his first clients.[86] Inoculators worked with groups of friends, with one inoculation leading to another. The aristocracy's quest for information was more intensive than extensive: it was less a matter of knowing the effect of the operation in general, than of obtaining reliable information about the practice of a specific inoculator, with regard to bodies with which one could identify. Personal medical accounts, rather than risk, provided socialites with the information they needed to convince them. We shall see how their social world produced this type of knowledge.

First, lay medical writing played an essential role. In contrast to the nineteenth century, when doctors confined patients' stories to oral accounts, in the mid-eighteenth century it was common practice for patients to report on their sick bodies, describe their symptoms, put it all down on a sheet of paper and send it to a friend or to their doctor.[87] The 'medical consultation by letters', then widespread in affluent circles, testifies to the recognition of patient narratives as reliable.

Since inoculation is a medical event *par excellence*, and one that had to be recorded, its history was often written by several hands: relatives and friends kept their own 'inoculation diaries', recording the condition of the patient and their pustules from one hour to the next. In the event of an accident, these diaries become formidable evidence against the

---

86   Among others: the children of Helvétius, of the Comte de Jaucourt, and of the Comte d'Houdetot. See the list in Angelo Gatti, *Lettre de M. Gatti, . . . à M. Roux, docteur régent de la Faculté de médecine de Paris, etc.*, 1763, pp. 4–6, which overlaps with the society described by Alan C. Kors, *D'Holbach's Coterie, an Enlightenment in Paris*, Princeton, NJ: Princeton University Press, 1976.

87   Wayne Wild, *Medicine-by-Post: The Changing Voice of Illness in Eighteenth-Century British Consultations Letters and Literature*, Amsterdam: Rodopi, 2006; Séverine Pilloud, 'Consulter par lettre au XVIIIe siècle', *Gesnerus*, 61, 2004, pp. 232–53; Vincent Barras and Philip Rieder, 'Corps et subjectivité à l'époque des Lumières', *Dix-Huitième siècle*, 37, 2005, pp. 211–33.

inoculator. To take an example, one of the rare accidents whose reality La Condamine conceded to the opponents of inoculation involved the children of the Marquis de La Perrière. The case had a considerable impact, prompting the publication of four booklets and several articles. In 1765, near Besançon, the two sons of the Marquis de La Perrière were inoculated by the doctor Acton. The operation went badly wrong: the younger son died, while the older son survived but was left deaf. The inoculator blamed a hereditary disease (syphilis was of course on everyone's mind) and the children's poor health. Outraged, the father wrote a 'history of the inoculation', which gave an extremely detailed account of the operation, with hardly any equivalent in medical literature:

> Mr A*** used scissors to cut their flesh at the elbow fat. He made a cut in each arm, removing the skin and a piece of flesh down to the blood, and put over each hole a thread of muslin soaked in the smallpox pus, a whole crust of a smallpox pellet, a small piece of muslin soaked and saturated in the said pus, a bit of cotton tied up like a pea and also saturated in the pus, and over all that a wad of dry lint and a plaster. The operation was so painful that my eldest son fainted, and the little one cried out.[88]

This memorandum was sent to prestigious English inoculators and to the editor of the *Journal Britannique*. They all criticised the inoculator's 'barbaric' method.[89] The Marquis de La Perrière also consulted François Dezoteux, an ambitious young surgeon who had just returned from London, where he had learned about inoculation. The misfortune that befell the marquis's son provided a perfect opportunity to bolster his reputation: on the pretext of 'reassuring the public', he published a pamphlet that criticised Acton and promoted his own method of inoculation as modern, English and safe. Generally speaking, accounts of accidents were often produced by inoculators themselves: inoculation was part of the competition between doctors for aristocratic patronage, and as the inoculation market was small but highly profitable, it was tempting to discredit competitors by publicising their accidents. The dangers of the practice were thus revealed by the (rather suicidal) rivalry between inoculators.[90]

---

88 François Dezoteux, *Pièces justificatives des lettres concernant l'inoculation*, Lons-le-Saunier, 1765.

89 François Dezoteux, *Lettres concernant l'inoculation*, Besançon: Charmet, 1765.

90 See for instance the conflict between the Nîmes doctors J. Nicolas (*Journal des inoculations*, Avignon: Chambeau, 1766) and J. Razoux (*Lettre sur l'inoculation de la petite vérole écrite de la vallée de Tempé*, Cologne, 1765).

Second, inoculation accidents fit in well with the news culture characteristic of the socialite world. Taking the form of short and entertaining
anecdotes, this news had to be surprising and unprecedented.[91] The
illnesses of the mighty so captured the public's attention that some periodicals dedicated a special section to them. *Le Journal de Paris*, for example,
offered to provide 'a bulletin on the illnesses of persons whose health is
of interest to the public'. In 1765, the Duchesse de Boufflers, who had
been inoculated by Gatti, suffered a recurrence of smallpox. This was a
huge story, reported in the *Journal des Dames*, the *Année Littéraire* and
the *Gazette Littéraire*: 'The whole of Paris was abuzz with the news.'[92]
According to Grimm: 'What has happened to Mme de Boufflers will make
a great noise in Europe.' The affair also occupied the salons' attentions.
The British writer Horace Walpole, living in Paris, claims that it was the
sole topic of conversation for a month.[93] This focus on the exceptional
deeply irritated propagandists for inoculation: what was the point of the
probabilistic argument based on the repetition of positive facts – which
were banal, to be sure, but also had a mass of numbers behind them – if
the slightest surprising development did more to capture high society's
attention?

Finally, the rules of politeness and sociability played a similar role
in publicising accidents. Medical accounts belonged to the epistolary
culture of the socialites, and many inoculation diaries and consultations
by letter circulated in the salons before they ended up on the pages of
some periodical. The reading aloud of medical correspondence also
contributed to the public presentation of bodies.[94] Inoculation redefined
sociabilities, rather than breaking them up. For example, in aristocratic
families, it was a common practice to choose loyal friends (who had
already had smallpox) to attend to the sick. These friends were quick to
tell of operation, in order to show off their devotion. To friends who were
less close or who had not themselves been immunised, the inoculated
person's family would usually send daily health bulletins. Simple acquaintances would send their servants, who would be given a note describing
the patient's condition. In society parlance, this was known as 'sending
someone to visit'. If the patient was a relative, it was customary to visit

91   Antoine Lilti, *Le Monde des salons*, Paris: Fayard, 2006, pp. 320–3.

92   Guillaume J. de L'Épine, *Supplément au rapport fait à la faculté de médecine
contre l'inoculation*, Paris: Quillau, 1767, p. 75.

93   *Horace Walpole's Correspondence*, New Haven, CT: Yale University Press, 1961,
vol. 31, p. 93.

94   Guillaume J. de L'Épine, *Rapport sur le fait de l'inoculation*, Paris: Quillau, 1767,
pp. 40–3.

or have someone visit them on a daily basis. The operation was usually concluded with a 'convalescence party' to praise the courage of the inoculated person and to show off the results – if, that is, they were flattering. Lastly, good manners required a 'convalescence visit', which offered the ideal opportunity to form an opinion on the practice. One could discuss the operation, find out about the pain it caused and examine the results with one's own eyes. For example, in 1764, when inoculation was still the exception in Besançon, socialites flocked to the door of the Marquis de Puricelli, who had had his daughter inoculated. Exasperated, the father agreed to show the arms of the convalescent girl so that the curious could inspect her scars.[95]

In short, the practices of writing and sociability around sick bodies and socialite news culture formed a highly effective system for monitoring the effects of the operation on a large number of bodies. This allowed for the extensive production of case studies, a constant watchful gaze (not a centralised and medicalised, but rather a profane and distributed one) and, ultimately, a fairly complete description of the consequences of the operation. As distinct from risk calculation, socialites could build a network of knowledge about the practice and its complications.

This socialite knowledge was highly valued by the doctors at the faculty. Indeed, to support their claims or authenticate a case, they would often invoke aristocratic authority. Socialites' word was all the more appreciated because, in accordance with the norms of noble civility, it had to stick to the truth, or else fall into dishonour.[96] Paris high society, through its collective experience of its members' health, thus became a medical authority in its own right: when the Paris *parlement* commissioned the faculty to produce a report on inoculation, the doctors' first initiative was to 'verify the facts that had transpired in the public, and were the subject of daily conversations'.[97] They interviewed inoculated patients and collected inoculation diaries and medical correspondence. In this way, academic expertise interacted with the knowledge of high society.

Thus, in the small world of the aristocrats who did choose to be inoculated, medical competence was widely distributed. The situation was

95   'Lettre de M. Puricelli à A***, 18 octobre 1764', *Réponse à une brochure intitulée Lettres concernant l'inoculation*, Besançon: Daclin, 1765, p. 33.

96   Steven Shapin, *A Social History of Truth: Civility and Science in Seventeenth-Century England*, Chicago: University of Chicago Press, 1994, pp. 3–125. Mme de Boufflers's testimony speaks for itself: 'A person of her rank and character can have no other motive than the truth, and the public will readily rely on her testimony' (*Gazette littéraire*, 6, 1765, p. 377).

97   De L'Épine, *Supplément au rapport*, 1767, p. 65.

not one of expert doctors versus passive patients, but rather one of individuals taking the initiative to carry out experiments on themselves or their children, and paying doctors with whom they had a relationship of patronage to do so. Patients participated fully in the operation: they discussed the different methods to be used with their relatives and doctors, wrote inoculation diaries, sent doctors to investigate accidents, ordered autopsies and even had experiments carried out. Thanks to a rich social fabric and the fine line between private bodies and their public representations, information on inoculation circulated fulsomely within this network. Through many little strokes, it painted a picture of the dangers of this innovation that was very different from inoculator's statistics.

Aristocratic society remained resistant to the project of risk-based governance. There was never a single, probabilistic approach to inoculation, but rather a multitude of ways of being in one's body. Putting on a show, demonstrating one's courage, thinking of oneself as exemplary and asserting self-control were legitimate ways of contemplating inoculation. Conversely, the fear of remorse, of moral infraction and of the distortion of filial relationships by a mercantile rationality represented respectable reasons for rejecting it.

The importance attached to the body and its representation produced an ever greater number of medical narratives, written records and oral descriptions, and increased their circulation. This process of generalised narration of bodies demonstrated how variable were the forms of smallpox and the effects of inoculation, and the degree to which the statistical knowledge about inoculation was a questionable and dogmatic simplification. The irreducibility of bodies and cases took precedence over risk. The consequences of inoculation were explored in a polyphonic public space, in which the distribution of knowledges and authority was the norm.

The programme for governing conduct by risk, as envisioned by mathematicians and philosophers, was showing its limits. Risk, that old tool of personal edification borrowed from the Protestant casuistry of the 1720s, was clearly insufficient to govern autonomous individuals in their relations with their bodies. To bring about the vaccinated society of the early nineteenth century, it would take technologies of proof and power of a wholly different kind.

2

The Philanthropic Virus

The French Revolution made life part and parcel of a new political project. By establishing the sovereignty of the nation (from the Latin *nascor*, to be born), it made *birth* the basis of authority: if the king had been God's minister on Earth, the revolutionary governments were, instead, the guardians of *national* sovereignty – that is, the sovereignty of a biological unit defined by birth and descent. Life was also the currency of the Republican pact: to take a share in sovereignty, one also had to be willing to sacrifice it. Rousseau called on citizens to take the following oath: 'I unite myself in body, in goods, and in will with the Nation . . . I swear to live and die for it.'[1]

The Republican nation was both a carnal and a political entity – conceived by the revolutionaries as an organism with the National Assembly as its head. The population was no longer simply a set of subjects. Rather, it constituted a body politic that could choose to improve its own performance. The fundamental revolutionary theme of 'national regeneration' included this labour of biological self-improvement.[2]

---

1   Cited in Pierre Nora's entry 'Nation' in François Furet and Mona Ozouf (eds), *Dictionnaire critique de la Révolution française*, Paris: Flammarion, 1988, p. 802.

2   Antoine de Baecque, *Le Corps de l'histoire, métaphores et politiques, 1770–1800*, Paris: Calmann-Lévy, 1993. It was doubtless Thomas Hobbes's *Leviathan* that anchored biological metaphor in political philosophy. See Roberto Esposito, *Communitas. The Origin and Destiny of Community*, Stanford, CA: Stanford University Press, 2010. On revolutionary biopolitics, see Dora Weiner, 'Le droit de l'homme à la santé, une belle idée devant l'Assemblée constituante, 1790–1791', *Clio medica*, 5, 1970, pp. 1209–23; Jacques Léonard, *La Médecine entre savoirs et pouvoirs*, Paris: Aubier, 1981, ch. 3; Mona Ozouf, 'Régénération', in Furet and Ozouf (eds), *Dictionnaire critique de la Révolution française*, pp. 821–31; Emma Spary, *Le Jardin de l'Utopie: l'histoire naturelle en France entre Ancien Régime et Révolution*, Paris: Muséum national d'histoire naturelle, 2005.

This politics of life granted new powers on bodies, and especially on those of the poor. This is exemplified by a decree of June 1793 which made smallpox inoculation compulsory for all children whose parents received public relief. This provision (which was unique in Europe; there is also no trace of its being applied) was presented to the National Assembly as the fruit of a political transaction. In exchange for their recognition of private property, the poor had the right to demand assistance from the nation. And, while the representatives of the national sovereignty waited for the Republic to mould enlightened citizens who would be able to recognise their own best interests, they had every right to impose the measures, such as inoculation, that would ensure the people's greatest happiness. As the deputy defending compulsory smallpox inoculation argued in 1793: 'Society must never lose sight of those who contract with it. It must take each individual at the moment of his birth, and not abandon him until the grave.'[3]

Lastly, this new republican power over life went hand in hand with the contemporary transformations in warfare. While in the eighteenth century war had been a circumscribed, recurrent and quasi normal phenomenon, during the Revolution and the Napoleonic Empire it was recast as the eschatological confrontation between one nation, its bodies and its vitality, and the coalition of European powers.[4] In this 'total war', the optimisation of life played an essential role.

It was in this particular context that in 1798, the English doctor Edward Jenner revealed the existence of a mysterious bovine disease, the cowpox, that immunised humans against smallpox. European states immediately took the innovation into their own hands. In the 1800s, 'vaccination' (from the Latin *vaca*, cow) was made compulsory in the armed forces of Britain, Prussia and France.[5] French Interior Minister Jean-Antoine Chaptal had only just established the system of prefectures when he entrusted it with the mission of vaccination as its top priority: 'No other subject more greatly calls for your attention; this is a matter of the dearest interests of the state, and of a sure method of growing the population.'[6] The vigour of

---

3   Intervention by Étienne Maignet, 'Commission des secours publics', *Archives parlementaires, recueil complet des débats législatifs de 1787 à 1860*, 26 June 1793, p. 493.

4   David A. Bell, *The First Total War: Napoleon's Europe and the Birth of Warfare as We Know It*, Boston, MA: Houghton Mifflin, 2007.

5   Peter Baldwin, *Contagion and the State*, Cambridge: Cambridge University Press, 1999, p. 235. Jean-François Coste, *De la Santé des troupes à la Grande Armée*, Strasbourg: Levrault, 1806. The low level of vaccinations in the Napoleonic armies owes to the fact that most of the soldiers had already had smallpox.

6   Circular of 14 germinal an XII (4 April 1804).

the first vaccination campaigns (at least 400,000 people were vaccinated in France in 1805, doubtless ten times the total number inoculated from 1760 to 1800) fit into a context of war mobilisation. The first vaccinators saw their action as analogous to war: their goal was to 'exterminate' small-pox or at least to 'extirpate' it from the national territory. Vaccination was presented as a 'conquest of art over nature',[7] as a Machiavellian stratagem using one virus to eradicate its peers.[8] Under Napoleon, the new form of inoculation was supposed to produce 'a magnificent race of men . . . able to compel respect for the [French] state abroad'. It would also painlessly accomplish that which the Revolution had attempted with fire and fury: the regeneration of man.[9]

Despite those grandiose aims, however, there was no general vaccine obligation in France until 1902. Napoléon himself rejected the vaccinators' urgent appeals.[10] His own power was supposed to reflect the power of the *patres familias* over his children, an authority the imperial regime did not wish to trample. Whereas Napoleon's Civil Code, contemporary to the cowpox vaccine, sought to restore paternal power, compulsory vaccination looked rather more like the revolutionary projects to weaken it.[11] In 1808, Interior Minister Joseph Fouché rebuffed a report by the vaccinators: 'The coercive measures which they propose are not at all authorised by law, and *gentleness* and *persuasion* are the most effective means of making the new inoculation a success.'[12] But how, under the Empire, could the government rule by *gentleness and persuasion*?

Vaccination is a classic object of historical inquiry. Historians have studied the organisation of vaccination campaigns, its spread across the

---

7   Gabriel Jouard, *Quelques Observations pratiques, importantes et curieuses sur la vaccine*, Paris: Delalain, 1803, p. 20.

8   Jacques-Louis Moreau, *Traité historique et pratique de la vaccine*, Paris: Bernard, 1801, p. 277.

9   J. Parfait, *Réflexions historiques et critiques sur les dangers de la variole naturelle, sur les différentes méthodes de traitement, sur les avantages de l'inoculation et les succès de la vaccine pour l'extinction de la variole*, Paris: Chez l'auteur, 1804, p. 67.

10   When a prefect proposed to make vaccination a prerequisite for baptism, he was informed: 'His Majesty has formally declared himself against severe measures' (AAM V 52, letter from the prefect of Landes to the central committee, 27 September 1811). As early as 1806, however, in the Principality of Piombino and in Erfurt, the French administration made vaccination compulsory for all children. Bavaria and Hesse followed suit in 1807. The obligation to vaccinate was established in Sweden in 1816 and in Great Britain in 1856. See Baldwin, *Contagion and the State*, pp. 254–66.

11   Jean Delumeau and Daniel Roche, *Histoire des pères et de la paternité*, Paris: Larousse, 1990, p. 279–312. The father 'supplements the laws, corrects morals and prepares obedience'; see P.-A. Fenet, *Recueil complet des travaux préparatoires du code civil*, Paris: Au dépôt, 1827, vol. 10, p. 486.

12   AN F8 97, *Rapport sur les vaccinations en France en 1806 et 1807*, p. 119.

globe, its demographic consequences, and the anti-vaccination movement. This chapter's focus is rather different: it refuses to consider the
cowpox vaccine as a fixed essence transmitted by neutral vectors. It instead
proposes a historical ontology of the vaccine, that is, a history of the
definition of its effects, of their representation and of the production of a
social consensus surrounding them. In 1800, the vaccine was little more
than a neologism. It was a mysterious entity with almost no existence, no
essence and as yet indeterminate capacities. The first vaccinators knew
perfectly well the failure of inoculation. They knew about the difficulty
of persuading people to risk their lives in order to protect them. Thus,
they set about fixing the characteristics of the vaccine in a way that could
stamp out any reluctance. They forced through the unlikely definition
of cowpox as a non-virulent virus, a *perfectly benign* virus, which would
protect people from smallpox *forever*. The government's strategy was not
to impose vaccination, but rather to establish and maintain a definition
of the vaccine such that anyone in their right mind would have to agree
to it. 'Transcendental legislation', Jeremy Bentham wrote 'leads men by
silken threads, entwined round their affections, and makes them its own
for ever.'[13] The vaccine of 1800 was one of the silken threads of power. In
establishing a new natural being, doctors sought to govern bodies not
through constraint, but indirectly, by orienting perceptions. The gentleness of power had, as its counterpart, its toughness in the domain of
evidence and truth.

1. The objectivity of the philanthropist

Doctors' image, their self-assuredness and the trust they commanded, were
decisive for the success of vaccination. *Le Bienfait de la vaccine* reflected
the image that doctors wanted to establish of themselves. On the left, a
peasant woman is taking her baby to a château so that a lady can have
her own child vaccinated. Alibert, a doctor at the Saint-Louis Hospital,
gently draws up the vaccine fluid from the cowpox pustule on the baby's
arm without even waking him up. The father keeps watch over the operation, while the mother tenderly reassures her child who is about to be
vaccinated. This painting depicts a peaceful society where peasants and
bourgeois live in harmony, brought together by the benefits of vaccination.
In the centre, Alibert, with his generous and intelligent visage, symbolises

13   Jeremy Bentham, 'Essays on the Subject of the Poor Laws', in Michael Quinn
(ed.), *Writings on the Poor Laws*, Oxford: Oxford University Press, 2001, p. 136.

Constant Desbordes, *Le Bienfait de la vaccine*[14]

the crucial place that medicine wished to occupy in post-revolutionary society, as the intermediary between the social classes. Whereas inoculation was an elitist practice, vaccination was an interface binding rich and poor: the state would take charge of the people's health, and in exchange for this people would make the transmission of the vaccine possible. The good health of all depended on the cooperation of each and every person, and enlightened philanthropy would relieve poverty not through charity but through innovation.

The philanthropic movement was the driving force behind the first vaccination campaigns. Even before the discovery of cowpox vaccination, philanthropists had hoped to generalise the practice of smallpox inoculation.[15] In England, parishes paid inoculators to immunise their poor populations. In Franche-Comté, doctors paid by the *intendant* (a provincial administrator under the Ancien Régime) inoculated children free of charge in the countryside.[16] These successes were, however, exceptional cases: everywhere else, and especially in the cities, the risks (of death and contagion) rendered unfeasible any general inoculation projects. For example, when doctors proposed to inoculate the foundlings of Lille free of charge, the hospital administration stressed the dangers of such an undertaking: these children's health was hanging

14   1822, Musée de l'APHP, Paris.

15   From the 1780s, the Philanthropic Society was a powerful institution bringing together financiers, intendants, aristocrats and scholars. See Catherine Duprat, *Pour L'Amour de l'humanité, le temps des philanthropes, la philanthropie parisienne des Lumières à la monarchie de Juillet*, Paris: C.T.H.S., 1993, vol. 1.

16   Peter Razell, *The Conquest of Smallpox*, London: Caliban, 1977; AD Doubs, 1 C 603; Pierre Darmon, *La Longue Traque de la variole*, Paris: Perrin, 1984, pp. 129–35.

by a thread, and inoculation could cause a return of the epidemic.[17] In 1790, La Rochefoucauld-Liancourt – a leading figure of the philanthropic movement and president of the National Assembly's beggars commission – proposed, without success, the establishment of a public inoculation hospital in Paris. Under the Convention, plans for compulsory inoculation were left entirely unimplemented. Finally, in 1799, barely a year before the arrival of the vaccine, inoculation was still being blamed for triggering an epidemic in Paris.[18]

Cowpox vaccination was thus of immediate interest to philanthropists: if defined in suitable terms – that is, as non-contagious and benign – it would ease the situation, by finally making possible their old project of generalised immunisation of the poor. On 15 February 1800, upon his return from emigration, La Rochefoucauld-Liancourt launched a public subscription in the *Journal de Paris* to finance a 'vaccine committee' charged with passing judgement on this innovation which was already causing a stir in Britain.

The financing of a 'vaccine committee' by prestigious notables (famous philanthropists, ministers, deputies, bankers . . .) removed this innovation from the market of remedies and created the impression of impartial assessment. The philanthropic movement allowed the vaccinators to fashion themselves a new and somewhat contradictory social role: that of the zealous but disinterested advocate of technology, both propagandist for an innovation and its judge. Philanthropy gave succour to the vaccinators' image of disinterestedness: they could speak as 'friends of humanity', promoting the cowpox vaccine as the greatest invention of the century, indeed an invention that would regenerate man, while presenting themselves as neutral, cool-headed judges, 'devoid of all sensibility'.[19] Forty years earlier, the advocates of inoculation had made no claim to stand above controversy. Then, the two sides had clashed and appealed to some judge – this could mean the faculty of medicine, a provincial *parlement*, or the public – but neither the propagandists nor their opponents claimed to occupy the judge's chair.

---

17   AAM V1, Observations of Sifflet, doctor of the poor of the parish of St Catherine, 1786.

18   *Journal de Paris*, 5 ventôse an VII (1 February 1799).

19   *Rapport du comité central de vaccine*, Paris: Veuve Richard, 1803, p. 10. This was an extraordinary thing to proclaim at the time: objectivity and self-effacement were not (yet) scientific virtues; on the contrary, sensibility and the ability to be moved by phenomena were the hallmarks of the empirical Enlightenment scholar. See Lorraine Daston and Peter Galison, *Objectivity*, New York: Zone Books, 2007, pp. 17–35; Jessica Riskin, *Science in the Age of Sensibility*, Chicago: University of Chicago Press, 2002.

To be philanthropic, the vaccine had to meet a set of precise criteria. First, it had to be perfectly benign. Inoculation had required the expertise of a doctor who selected and prepared the subject; the cowpox vaccine, conversely, was presented as absolutely safe, regardless of the patient's state of health. This was the condition on which it could be carried out by simple health officer, midwives or even priests. Philanthropy sought vaccination for all, rich and poor, healthy and feeble, newborns and the bedridden; it thus had to produce a risk-free vaccine.[20]

Second, the vaccine would have to be free. The vaccines committee thus systematically rejected doctors who tried to make money from innovation. For in promoting their pus as more effective, safer or coming directly from London, they implied that the different qualities of pus existed, which risked discrediting everyone's vaccines.[21] The problem was that, in rejecting these mercantile but justified ventures, the committee held back the recognition of risks and the search for safer vaccine techniques.

2. Uncertainties and catastrophism

When, in the spring of 1800, a new inoculation called cowpox began to make news in Paris, doctors had some reason to be sceptical.[22] What was the nature of this pus? Could one even be sure that one was inoculating the real cowpox? And how was it possible to answer this question, when the material imported from England had passed via hundreds of bodies susceptible to various ailments?

In May 1800, the committee sent for the vaccine from London. Dr William Woodville, who practised vaccination at the Smallpox Hospital, sent pus in a hydrogen-filled vial, sealed with mercury, covered with a bladder and placed in a padded box.[23] A month later, the committee

---

20   Henri-Marie Husson, *Recherches historiques et médicales sur la vaccine*, Paris: Gabon, 1801, p. 50.

21   Dr Colon, one of the main promoters of the vaccine, in whose house the committee conducted its first experiments, was thus excluded from the committee for having tried to sell the vaccine. See *Moniteur Universel*, 28 vendémiaire an IX (20 October 1800).

22   L. A. Mongenot, *De La Vaccine considérée comme antidote de la petite vérole*, Paris: Méquignon, 1802, p. 4: 'For some time it was regarded as a brilliant chimera or as pure charlatanism. I must confess that at first I could not escape this unfavourable impression.'

23   AAM V6 d54, letter from Alexander Pearson and Woodville to Thouret, 6 May 1800. In January 1800, Pinel had unsuccessfully attempted to inoculate a pus taken from cows in Paris. After this failure, it was generally accepted that cowpox did not exist in French cowsheds.

diluted the dried-out pus and inoculated thirty foundlings. Nothing worked as expected: only seven children had a pimple; a little boy called Blondeau who had the best-looking vaccine pustule was then inoculated with smallpox and caught the disease. The Parisian vaccinators groped around for explanations: the rashes they obtained showed no regularity and they could not identify cowpox. To put an end to the uncertainties, the committee invited Woodville himself to Paris. In October 1800, he vaccinated 150 children.[24] The controversy now began.

According to some doctors, the cowpox Woodville had brought was, in fact, a smallpox. Indeed, during the first vaccinations at the Smallpox Hospital in London, most patients had developed a generalised rash rather than a vaccine pustule. Woodville himself had initially speculated that Jenner's pus was merely smallpox, before admitting that the atmosphere in the hospital might have caused the epidemic among his vaccinees.[25] In fact, cowpox was initially perceived as a mild form of smallpox. In the late eighteenth century, doctors had attempted to mitigate the variolic pus by bathing smallpox scabs in hot water, weakened acid or various gases.[26] It seemed that Jenner's achievement was simply to have discovered a way to attenuate smallpox with the help of the cow.

But by October 1800 the vaccine's symptoms appeared to have nothing in common with those of smallpox, and doctors agreed that it was in fact an entirely different disease. The problem was that, if the vaccine was not ultimately a mild form of smallpox, then what was it – and more importantly, what could explain its effectiveness? The vaccine system was easy enough to criticise from a theoretical point of view. How could a local, almost feverless movement destroy the constitutional trait that disposed the individual to smallpox? How could smallpox occur *during* cowpox? Strangely, the cowpox vaccine would appear to protect against all future smallpoxes but not against the present one. Opponents used the chemical analogies then commonplace in medicine to highlight the absurdity of this delayed effect: it was as if the vaccine was powerless to neutralise smallpox when in contact with it, but then became effective once it was absent from the body. Inoculation was tantamount to anticipating

---

24   'Rapport du comité médical établi à Paris pour l'inoculation de la vaccine', *Moniteur universel*, 28 vendémiaire an IX (20 October 1800).

25   This episode is the source of a long dispute among historians. Peter Razzell has taken up the thesis of the contamination of the Woodville vaccine by smallpox, and thus denounces the 'myth' of Jenner's cowpox. See Peter Razzell, *Edward Jenner's Cowpox Vaccine: The History of a Medical Myth*, Lewes: Caliban Books, 1977.

26   M. Bouteille, 'Troisième dissertation sur l'inoculation', *Journal de médecine*, 47, 1777, p. 226.

a natural phenomenon by causing smallpox; inoculating an unknown material seemed a much more reckless undertaking.

The problem with the cowpox was not so much that it was risky, in the sense that the inoculation carried a risk of death for the individual, but rather that the nature of its potential dangers was unknown. Unlike smallpox, cowpox was rare and was not contagious. It therefore had to be transmitted from arm to arm, from vaccinators to vaccinated, in an ever-extended chain. Vaccinating thus amounted to inoculating a virus that had flared up in hundreds of bodies that might be affected by all manner of diseases. Because it could transmit syphilis or scrofula, which were hereditary diseases, cowpox put at risk the health of 'all future generations' or even 'the constitution of the human race'.[27] By hybridising, cowpox could also create new diseases: 'Viruses . . . blend with each other and form even more formidable compound viruses: they spread by generation as well as by contagion. They degrade the national temperaments.'[28]

The possibility of disaster offered a solid reason to continue with experiments before spreading a new virus among the population. The German physician Marcus Herz thus emphasised the need to postpone any general use of the vaccine. The smallpox control tests on hundreds of vaccinated people, and the good health of thousands of others, proved nothing: the question was not the number of experiments but their duration: '50,000 trials are not enough to complete the experiment, and 100,000 would not prove anything further.' The problem was judging the longer-term consequences of this innovation. First, then, it was necessary to stop using cowpox inoculations, and instead pay close attention to the fate of those who had already been vaccinated. After ten years, the results could be communicated to the public and to doctors for discussion. If it seemed an obvious success, then another 50,000 individuals could be subjected to the same procedure. If, after a generation, the vaccine was still in good standing, it could finally be propagated across the entire population. This was the only way of acting with the rigour demanded by the colossal scale of what was at stake: the health of the European population and of the future generations.[29]

---

27 François-Ignace Goetz, *De L'Inutilité et des dangers de la vaccine prouvés par les faits*, Paris: Petit, 1802, p. 87. Jean-Sébastien Vaume, *Les Dangers de la vaccine*, Paris: Giguet, 1801, p. 48.

28 Jean Verdier, *Tableaux analytiques et critiques de la vaccine et de la vaccination*, Paris: Chez l'auteur, 1801.

29 Marcus Herz, 'Über die Brutalimpfung und deren Vergleichung mit der humanen', *Hufeland's Journal der praktischen Heilkunde*, 12, 1801, p. 3.

3. Test-tube bodies

At the origins of vaccination was a profound change in the role of medical experimentation on humans. Defining this change is tricky because human experimentation is not a category with clear-cut boundaries: the art of clinical evidence is precisely to attach experimentation to a therapeutic project in order to bring it closer to simple observation. But if there is a continuum between the observation of trials with a therapeutic aim and human experimentation with a purely evidential objective, then the trials carried out by the National Vaccine Committee in Paris between 1800 and 1803 were atypical – for they subjected a large number of children to experiments that had no therapeutic aim whatsoever. In this particular endeavour, medicine acquired a wide latitude in its use of bodies.

In the eighteenth century, the experimental subjects had been either doctors themselves or individuals who had been condemned to death.[30] In this case, experiments on humans were morally no different from dissection, since the bodies subjected to the experiment were legally already dead. Being dependent on the sovereign's power to punish, human experimentation was not a normal method of gathering medical proofs. During the dispute over inoculation, the few cases of experimentation on humans were not the work of doctors but of aristocrats who had their servants' children inoculated with pus to test its goodness before administering it to their own offspring.[31] But the cowpox vaccine broke out of these restrictive frameworks. The number of bodies subjected to experimentation rose by orders of magnitude: no longer just a matter of a few men on death row, but hundreds and hundreds of abandoned children.

In 1800, the vaccine was something new. What it was and what it could do were uncertain, and many vaccinations went wrong.[32] The vaccine was also rare and fragile: its existence depended on its transmission; and if doctors had no more subjects to vaccinate, it would disappear. Attempts to preserve the pus *ex vivo* – in glass plates, capillary tubes, vacuum flasks

---

30   Grégoire Chamayou, *Les Corps vils. Expérimenter Sur Les Corps humains aux XVIIIe et XIXe siècles*, Paris: La Découverte, 2008, pp. 21–94. Abandoned children were sometimes used to demonstrate the usefulness of remedies, but the doctors were usually rebuffed by the foundling hospital administrators.

31   De l'Épine, *Supplément au rapport*, p. 120.

32   Pierre Chappon, *Traité historique des dangers de la vaccine, suivi d'observations et de réflexions sur le rapport du comité central de vaccine*, Paris: Demonville, 1803, lists 207 cases of smallpox after vaccination, 39 deaths and 115 accidents of various kinds. In AAM V30 d2, we find letters addressed to Husson recounting accidents and cases of recurrent disease.

or vials filled with nitrogen – were fruitless. Vaccines were much more successful when carried out with fresh material, passed from arm to arm. So, to preserve and transport this protective virus, doctors had to organise vaccination chains. Throughout the nineteenth century, foundlings provided the indispensable links in this chain.

Let us take an example. In October 1800, Dr Husson, secretary and lynchpin of the National vaccine committee, arrived with the precious lymph in Reims, where he inoculated his family and friends. Once back in Paris, he published an optimistic article: 'The Fire of the Cowpox Vaccine Is Kept Burning.'[33] The metaphor, taken in a prehistoric sense, was judicious: the first vaccinators struggled to maintain lasting chains of transmission. They were forever looking for children to vaccinate in order to maintain the virus; but few parents were willing to surrender their children to the lancet, especially as winter approached.[34] The doctors in Reims had an imperative need to vaccinate the foundlings, which the hospice administration refused. In his private correspondence, Husson expressed his doubts: 'I fear that the vaccine will soon fade away.'[35]

Chaptal's appointment as interior minister on 7 November 1800 changed the political situation surrounding the vaccine. Pinel, whom he had met in the 1770s in the Montpellier medicine faculty, convinced him that the innovation would make it possible to eradicate smallpox.[36] From 1801, on his orders, the hospices were opened up to vaccinators. The power balance had changed. When a prefect refused them access to foundlings, the vaccinators replied, threatening: 'It would be painful for us to be forced to write about this to the interior minister, who is our associate.'[37]

This was the precise moment when the vaccine's future was assured: foundlings would be used to produce and transport the lymph until the end of the nineteenth century. As Yves-Marie Bercé has pointed out, 'without them, nothing would have been possible.'[38] A decree from 1809 designated twenty-five hospices for foundlings, which, under the

---

33   *Journal de médecine*, vol. 1, vendémiaire an IX, 1800, p. 266.

34   AAM V 57, Caqué to Husson, 21 nivôse an IX (11 January 1801).

35   Ibid. Husson to Caqué, 6 nivôse an IX (6 January 1801).

36   Jean-Antoine Chaptal, *Mes Souvenirs sur Napoléon*, Paris: Plon, 1893, p. 19; Jean Pigeire, *La Vie et l'œuvre de Chaptal (1756-1832)*, Paris: Domat, 1931. AAM V1, the committee to Chaptal, 15 frimaire an IX (6 December 1800).

37   Ibid. Letter from Augalnier, doctor at the hospital in Marseille, to Thouret, 18 germinal an IX (8 April 1801).

38   Yves-Marie Bercé, *Le Chaudron et la lancette*, Paris: Presses de la Renaissance, 1984, p. 70.

euphemism of 'vaccine depots', were responsible for maintaining the vaccine. Since administrators and nuns were often reluctant to vaccinate vulnerable newborns, authority was instead entrusted to a doctor chosen by the prefect for his vaccinating zeal.

Maintaining the vaccine was a delicate job: the vaccinations had to be spaced out in order to provide fresh fluid on demand, and the 'vaccine depot' child had to be inoculated repeatedly in order to produce more pus (eight times in the Paris vaccine hospice, sometimes up to fifty times when not enough children were found).[39] To extract the precious fluid, the pustules were opened and then pressed several times over. This operation was carried out in public: in the villages, the mayor would be warned of the vaccinator's impending arrival, and then had to be present with the unvaccinated children in the town hall. This saved the vaccinator wasted journeys and allowed the vaccines to be put on a proper legal footing: the certificates (necessary for school enrolment and obtaining public relief) were often signed by the mayor.[40] The biological instrumental-isation of poverty did not itself cause alarm among parents. Above all, they were suspicious of the abandoned children's health: they feared that these foundlings, the product of depravity, would transmit syphilis; they demanded to be able to inspect their bodies, and preferred the pus of legitimate children. They reproached the system for being dangerous, but never for being inhumane.

Children in hospices were also used as a testing ground, where vacci-nators gained the skills and experience necessary to judge good vaccines from bad. Since accidents could easily be hushed up, vaccinators did not run the risk of parental recriminations or giving grist to the mill of anti-vaccine polemics. At the beginning, it should be remembered, doctors knew nothing about the cowpox vaccine. They needed to identify its characteristics and learn to reproduce them reliably. The committee estab-lished, for example, the time necessary for vaccination to give effective protection by organising smallpox control experiments on forty children: first, the vaccine and smallpox were injected at the same time, and then the second inoculation was delayed day after day. Another problem: at what age could children be vaccinated? The committee operated on younger and younger children, even vaccinating premature babies. No age seemed unsuitable. It might also be that the protection afforded by the vaccine

---

39   AN F8125, State of the individuals vaccinated by the undersigned, Bernardin Piana, 1812.

40   AN F8100, Circular from the prefect to the mayors, 15 July 1811.

was only local. Smallpox was thus inoculated at the extremities opposite the points of vaccination. Vaccinators also tried to reproduce accidents: they put pus in the throat or on the nasal mucous membrane in order to study the respiratory complications connected to the vaccine. Similarly, to understand vaccine rashes, they would take off the skin locally and apply a few drops of vaccine, resulting in a gangrenous wound. Inside the hospices, human experimentation became commonplace. Foundlings were used as test-bodies: if a previously vaccinated person developed a rash disease, the committee inoculated a hospice child with the offending pus to check that it was not smallpox.[41]

Trials were also run with the aim of refuting claims that the cowpox vaccine risked spreading infections: Alibert vaccinated subjects with leprosy, scrofula and ringworm by passing the pus from one to the other, without observing any cross-contamination. Cullerier and Richerand transferred vaccine taken from syphilitic children to healthy children without infecting them. These dangerous and much criticised experiments were carried out on only a few children.[42] However, they had important consequences, for throughout the nineteenth century, after hypothetical cases of syphilitic contamination, doctors seeking to clear the cowpox vaccine of blame invoked the great experiments that the committee had supposedly carried out in the early 1800s. In short, thanks to experimentation on foundlings, the committee explored and defined the competencies of the vaccine. The philanthropic programme of a virus that could be inoculated into everyone began to take shape.

The question here is not whether the vaccinators behaved inhumanely, but rather what legal and ideological mechanisms allowed them to do so. Hospice children were not simply 'vile bodies' on which anything was permitted. In fact, many doctors were outraged by the first vaccinators' trials.[43] In April 1800, the administrators of the hospices and then the interior minister had refused the Vaccine Committee the foundlings

---

41   *Rapport du comité central de vaccine*, pp. 97, 117, 203, 230–59.

42   P.-L. Delaloubie, *Essai sur l'emploi du fluide vaccin pris sur une personne atteinte de maladie ou de vice héréditaire, ou d'affection quelconque. Ce Fluide peut-il être nuisible ou sans danger?*, Dissertation in the Paris Faculty of Medicine, 1805. Richerand, often cited as having proven the non-transmissibility of diseases, seems to have carried out only one experiment of this type. Anthelme Richerand, 'Observations sur la vaccine', *Journal de médecine*, 2, an IX, p. 114.

43   When the committee reported on a trial attempting to transmit ringworm by vaccine, Dr Chappon was outraged. It was a 'cruel experiment'. He lost '[his] cool so much [his] sensibility was affected'. Chappon, *Traité historique des dangers de la vaccin*, pp. 279–81.

it had requested for its first experiments.[44] In Nantes, the Daughters of Charity who witnessed these efforts took exception to them.[45] For Marcus Herz, the vaccinators' experiments were unacceptable: the effect of the cowpox vaccine should be justifiable by analogous reasoning, or it should at least be possible to increase the dose little by little, or else experiment on subjects who had nothing more to lose.[46]

Dr Duchanoy, a member of the hospice administration, proposed a simple legal solution to the interior minister: it would be sufficient to place the doctors convinced of the benefits of the vaccine in the hospices. Doctors 'are in fact the natural judges to consult in this matter', and the experiments would be legal so long as they expected them to be of therapeutic benefit to the patient. Human experimentation had lost the clarity of its earlier moral contours; whereas it had previously rested on the sovereign's right to decide life and death, now it became the prerogative of the physician.[47]

The transformation of hospices into 'vaccine depots' and the use of foundlings as biological instruments were part of the project of making public relief more profitable. Historians addressing the treatment of these abandoned children have described a late-eighteenth-century transition from a charitable logic to a utilitarian one. This immense population (60,000 children in 1800), its miserable fate and its potential usefulness to society made it the object *par excellence* of a philanthropic practice seeking both to relieve destitution and to find it a social function.[48] In this same moment, the French navy secured the right to enrol the children under the state's stewardship because, 'having been brought up in the charge of the state, they *belong to it*'.[49] Compared to Trafalgar, the vaccine hospice seemed a lesser evil.

---

44   AN F8 97, The members of the medical committee of the vaccine inoculation society formed in Paris to the interior minister, ventôse an VIII.

45   AD Loire Atlantique, 1M 1344, letter from the vaccination committee to the prefect of Loire-Inférieure, 21 February 1807.

46   Herz, 'Über die Brutalimpfung'; AAM V1, 'L'Inoculation brutale'.

47   AN F8 97, Duchanoy to Lucien Bonaparte, 29 germinal an VIII (19 April 1800).

48   Jehanne Charpentier, *Le Droit de l'enfance abandonnée*, Presses Universitaires de Rennes, 1967; Muriel Joerger, 'Enfant trouvé-enfant objet', *Histoire, économie, société*, 6, 1987, pp. 373–86. Before vaccination, artificial breastfeeding was the great mass experiment of the eighteenth century. See Marie-France Morel, 'À quoi servent les enfants trouvés? Les médecins et le problème de l'abandon dans la France au XVIIIe siècle' in Jean-Pierre Bardet (ed.), *Enfance abandonnée et société en Europe, XIVe–XXe siècles*, Rome: École française de Rome, 1991.

49   Law of 15 pluviôse an XIII (3 février 1805). See A. and D. Dalloz, *Répertoire méthodique et alphabétique de législation de doctrine et jurisprudence*, Paris: Bureau de la jurisprudence générale, 1855, vol. 32, entry 'Minorité, tutelle, émancipation', p. 241.

4. What did the control tests prove?

In the Eighteenth Century, the difficulty of acclimatising experimental proof in medicine was statistical in nature: unlike physical phenomena, medical causalities were always in flux, and it did not seem very rigorous to prove the effectiveness of a therapy just using the few experimental subjects the sovereign granted.[50] The availability of children in hospices allowed vaccinators to overcome this obstacle and to import the epistemology and discourse of the crucial experiment into medicine.

In November 1801, the Vaccine Committee organised a large-scale public experiment in Paris: 102 foundlings, who had previously been vaccinated, were inoculated. Various notables were invited to testify to the success of the experiment, that is, to confirm the absence of smallpox. According to the Académie des sciences, 'the result is the most decisive experimental proof one could ever desire'.[51] The British Parliament was also very impressed: Jenner had carried out control tests, but on only four subjects.[52] The probabilistic dimension of bodily phenomena disappeared with the discourse of the crucial experiment: 'It is not a question of determining the degree of probability, but rather the infallibility of the new mode: either it protects from smallpox or it does not.'[53] Throughout France, the same administrative ceremony of experimental medical proof was repeated. By 1803, the Vaccine Committee's bet had paid off: the interior minister, the prefects and the notables were all convinced by the control tests.

Yet, at the same time, some doctors remained sceptical. First, even if the cowpox vaccine protected against the inoculation, it might not protect against natural smallpox when the virus was breathed or ingested. Second, inoculations were carried out no later than eighteen months after the vaccine, but the real problem was how long the protection would last. Third, children in hospices provided convenient material to be

50   François Doublet, entry 'Expérience particulière' in *Encyclopédie méthodique. Médecine*, Paris: Panckoucke, 1793, vol. 6, p. 180.

51   Antoine Portal, Jean-Noël Hallé, Antoine-François Fourcroy, Jean-Baptiste Huzard, *Rapport fait au nom de la commission nommée par la classe des sciences mathématiques et physique, pour l'examen de la méthode de préserver de la petite vérole par l'inoculation de la vaccine*, Paris: Baudoin, germinal an XI (1803).

52   Jenner gave the British Parliament a presentation on the Vaccine Committee's experiments in order to make the case for the benefits of his discovery. See *The Evidence at Large as Laid Before the Committee of the House of Commons, Respecting Dr Jenner's Discovery of Vaccine Inoculation*, London: Murray, 1805, pp. 172–4.

53   Louis Valentin, *Résultats de l'inoculation de la vaccine dans les départements de la Meurthe, de la Meuse, des Vosges et du Haut-Rhin*, Nancy: Haener et Delahaye, 1802, p. 32.

experimented upon, but also a defective material, for their lack of parents meant their medical histories were unknown: their immunity might owe not to the vaccine but to some unknown case of smallpox.[54] Finally, even if smallpox was an externally visible disease, diagnosing it was not as easy as many thought. Deciding whether a child had had smallpox was a delicate matter, for which the administrators invited as witnesses had no credentials. Since the purpose of the smallpox inoculation was precisely to avoid a general rash and to reduce smallpox to local symptoms, the absence of a general rash in the 102 experimental subjects was not itself sufficient proof.[55] The real question was much more nuanced: was there a difference in nature between the local symptoms of post-vaccine inoculation, and those of a simple inoculation?

When the control test was subjected to an expert critical eye, it lost its decisive character and turned into yet another problem to be solved. It was no longer enough to attend and witness it – one had to know how to read it properly, to know what was a local reaction as distinct from a general one, and to know the symptoms of an inoculation in individuals who had or had not had smallpox before. Thus, the criteria that defined the success of the test were themselves the object of controversy.[56] The corollary was that, the more convincing was the test – insofar as the witness lacked real medical knowledge of pustular eruptions – the control tests were particularly persuasive for the administrators.

As a result of this controversy, the capacities that philanthropists attributed to the new virus were translated into more ambiguous terms. The proposition *the cowpox protects against smallpox* became *there is no eruption after a smallpox control test*, which became *there is no variolous progress at the incisions*, which, in turn, was translated into *there is a difference in nature between the progress of a smallpox inoculation and a post-vaccine smallpox inoculation*. In their attempts to qualify this difference, vaccinators now embarked upon a pioneering enterprise of mapping the ever moving and contested world of pustular eruptions.

---

54   François-Ignace Goetz, 'Au rédacteur', *Le Moniteur universel*, 10 brumaire an IX (1 November 1800).

55   François-Ignace Goetz, *Traité de la petite vérole et de l'inoculation*, Paris: Croullebois, 1798, p. 59; M. A. Salmade, *Instruction sur la pratique de l'inoculation de la petite vérole*, Paris: Merlin, 1799, p. 55.

56   One never tests a hypothesis in isolation, but a web of non-explicit hypotheses or beliefs. See Harry Collins, *Changing Order: Replication and Induction in Scientific Practice*, Chicago: University of Chicago Press, 1985.

THE PHILANTHROPIC VIRUS   63

5. The graphic nature of the vaccine's power

The story begins with a particularly devious footnote on page seven of Jenner's *Inquiry*.[57] Before he presented the cases which supported his theory – stories of cowpox victims who were thereby protected from smallpox – Jenner referred to a mysterious bovine disease closely resembling cowpox, and transmissible to humans, but which did not immunise against smallpox: *spurious* cowpox. This distinction gave Jenner's theory a degree of protection: the numerous cases of post-vaccine smallpox reported to him by his colleagues could be attributed to the false cowpox without damaging the reputation of the real thing. The notion of false cowpox, initially marginal enough to be relegated to a footnote, soon became the central cog in the vaccine theory. For its opponents, this was a crude subterfuge designed to render the theory irrefutable. It was on this precise point that the new clinical gaze played a decisive role: instead of explaining false cowpox in terms of the causes of its appearance (causes that became more extensive and more mutually contradictory as more and more post-vaccine smallpoxes appeared), the Vaccine Committee established, through the clinic, the typical pustule of the *true cowpox* and rejected all those that did not meet this definition. This is not to say that the committee or Jenner himself were playing tricks: there are, of course, vaccinations that do not protect. Rather, what was at issue was the definition of the criteria for the success of the vaccine and, by the same token, those who were capable of judging it.

In the treatises on the cowpox vaccine from the beginning of the century, dozens of pages were devoted to extraordinarily detailed descriptions of the pustule: its size, shape, colours, consistency, relief, induration and elasticity, across the different stages of its existence. This minute detail was itself new. In the eighteenth century, since symptoms were thought of as simple signs of the disease, the essence of which remained inaccessible, to devote oneself to their exhaustive description was tantamount to chasing shadows. Inoculation, in its very principle, was based on this disconnection: one could hope to have smallpox without the rash, the disease without its symptoms. The novelty of the clinical gaze lay in its erasure of the separation between disease and symptom: there was no longer a pathological essence, and the disease was nothing more than the collection of symptoms in which it materialised.[58] So, when the doctors

---

57   Edward Jenner, *An Inquiry into the Causes and Effects of the Variolae Vaccinae*, London: Sampson Low, 1798, p. 7.

58   Michel Foucault, *The Birth of the Clinic: An Archaeology of Medical Perception*, London: Routledge, 1973, pp. 88–106.

of 1800 observed the vaccinal pustule, they saw the disease itself. The pustule was not the *sign* of cowpox; *it was cowpox*. Its clinical description allowed it to be classified among the eruptive diseases:

> In the eyes of educated men, cowpox will always be totally distinguished
> by its great natural characteristics, namely:
> 1. the central depression,
> 2. the areola,
> 3. the subcutaneous tumour,
> 4. the clarity of the fluid,
> 5. its deposit in isolated lodges or cells,
> 6. the silver tint of the pustule, and
> 7. finally, its very regular shape.[59]

The clinical definition also made it possible to justify failed cowpox: if a pustule did not develop this series of visible phenomena, then it was not a true vaccine and could not therefore protect against smallpox.[60] This also explains the committee's refusal of booster vaccinations up until the 1840s: since it was difficult to obtain a 'true cowpox' pustule during a second vaccination, and, given that this pustule was the criterion of its protective effect, the committee insisted on the importance of a good first vaccination and rejected the abnormal pustules produced by second vaccinations.

The problem with the clinical definition was that it was based on nuances that were themselves hard to describe. The colour palette of treatises on cowpox is particularly rich: 'light red', 'greyish white', 'opaline colour', 'pink tint', 'slightly purple tint', 'yellowish', 'tan colour similar to barley sugar', 'mahogany wood', and so on.[61] Colours, which had been considered secondary qualities by natural philosophy, botanical systematics and medical nosology, became in these clinical descriptions the essential criterion of the most essential disease of the new century.[62] For instance, Husson insisted upon the 'inflammatory areola of a bright red colour with a white glaze'; the 1803 report makes the silvery colour of the bulge the determining criterion of true cowpox, and compares its colour 'to that of the fingernail whose tip is pressed'.[63]

---

59   *Rapport du comité central de vaccine*, 1803, p. 77.
60   Ibid., pp. 190–230.
61   Husson, *Recherches historiques et médicales*, pp. 32–6.
62   Linnaeus urged botanists to consider only form, proportion, number and position as characteristics. See Michel Foucault, *The Order of Things: An Archaeology of the Human Sciences*, London: Routledge, 1966, pp. 144–53.
63   Henri-Marie Husson, 'Réflexions sur la vaccine', *Recueil périodique de la Société*

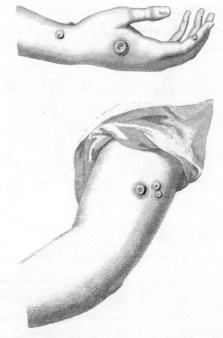

Edward Jenner, *An Inquiry Into the Causes and Effects of the Variolae Vaccinae*, London: Sampson Low, 1798

This is where the vaccinators' great innovation comes in: the *graphic definition of cowpox*. From the beginning of its work, the committee called on the services of Anicet Lemonnier, a painter and draughtsman attached to the School of Medicine.[64] His involvement is often mentioned: 'As he had so many opportunities to follow cowpox in all its nuances, varieties and degenerations, he drew on the help of drawing and painting to transmit faithful images of its development on man and on the cow . . . and in its various states of true or false cowpox.'[65] The rapid progress of the representations of the pustule shows that vaccinators were indeed inventing a new graphic code.

The sociology of science has shown how, by selecting what is to be perceived, scientific images augment the visibility of nature and define what thereby becomes knowable.[66] In 1800, the simultaneous presence

---

de médecine de Paris, 10, 1800, p. 118; *Rapport du comité central de vaccine*, 1803, p. 67.

64   Anicet Lemonnier (1743–1824) was a renowned historical painter. He was awarded the Prix de Rome and became a member of the Académie Royale de peinture in 1789. In 1794 he was appointed painter-draughtsman of the Paris École de médecine.

65   *Rapport du comité central de vaccine*, 1803, p. 26.

66   Michael Lynch, 'Discipline and the Material Form of Images: An Analysis of Scientific Visibility', *Social Studies of Science*, 15, 1985, pp. 37–66.

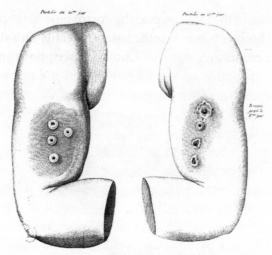

Hugues Félix Ranque, *Theory and Practice of Vaccinal Inoculation*, Paris: Mèquinon, 1801

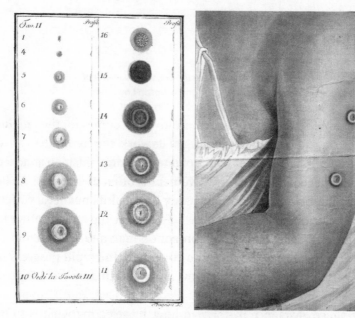

AN F8 97 and Henri-Marie Husson, *Historical and Medical Research on the Vaccine*, Paris: Gabon, 1801

of the painter and the clinician around the same cowpox pustules helped stabilise the phenomena, standardise observations and make the pustule a docile object of knowledge. The vaccinators' rich palette was the result of their collaboration with painters. The nosology of the cowpox vaccine, which transformed nuances into essences and colours into categories, thrived on the richness of the pictorial lexicon. By choosing colours, shades, reliefs and textures, the painter and the clinician together defined the characteristics of the true cowpox: the doctor guided the painter's gaze and controlled his brush, whereas the latter set reality on a stable footing and helped the clinician to name it. Clinical descriptions and images of disease were reciprocally constituted: the clinical and graphic definition of the true cowpox was constructed by lexical transfers from painting to the clinic, in a strange cognitive space constituted by a clinical experience restructured by the image.

By studying the composition and use of these images, we gain insight into the heart of medical power – that is, where this power was exercised in silence, in a moment when language had not yet given things settled names. The vaccinal image is not a copy of nature, but an interpretation based on nature; it is not a still-life rendering of the particular appearance of a pustule, but a typical image accentuating the characteristics that define cowpox. Doctors corrected the artist when he reproduced insignificant details. The *typical* cowpox was also understood as the *finest* specimen, which magnified the regularities discovered through clinical experience. Dr Fournier had no hesitation in choosing which model to have engraved on the basis of aesthetic criteria: 'Among more than four hundred individuals of both sexes that I have vaccinated [was] a girl of six years of age of an accomplished beauty . . . who gave me the most brilliant areola that I have yet seen.'[67]

The details deemed essential were described with extraordinary precision. Indeed, the images reproduced above display an armada of details. The colours were essential and so, too, was time. Hence the *serial* representation of the cowpox: it was not enough to develop a lesion that somehow resembled the characteristic pustule; it was also necessary that the lesion displayed the same succession of phenomena over time. We might also note the *cross-sectional* representation of the pustule which makes it possible to define the cowpox by touch. In postulating the visibility of the pathological and refining the medical gaze, clinical experimentation allowed the vaccinators to transform the detail into a particular sign – into a lapsus of nature, which betrayed itself in the fold of a pustule. In establishing an equivalence between the optimal subject and fidelity to nature, the culture of medical representation produced an effect of power. For, in purifying, synthesising and amplifying certain vaccinal phenomena, it transformed the different cowpoxes into so many varieties of false, abnormal, rachitic and irregular cowpoxes.

---

67   François Fournier, *Essai historique et pratique sur l'inoculation de la vaccine*, Brussels: Flon; Paris: Croullebois, 1802, pp. 14–19.

The nosologies designed to protect cowpox from criticism were refined over time. In addition to false cowpox, in the reference dermatological atlas of the 1830–1840s Pierre Rayer proposed the category of the 'vaccinelle': the virus had taken hold, the pustule showed the characteristic signs (the central depression in particular), but it went through the various phases of its development much more quickly than true cowpox. And, of course, it did not protect against smallpox.[68]

The same nosological confusion occurred further along the order of reasoning, with regard to smallpox itself. Since the eighteenth century, a distinction had been drawn between chickenpox and smallpox, each of which produced diseases of varying severity. From the 1810s onwards, during smallpox epidemics, many of the vaccinated displayed symptoms similar to smallpox, even if they were somewhat tempered. If the proponents of the cowpox vaccine were to defend its reputation, then these diseases simply could not be smallpox. So, it was necessary to find a disease similar to smallpox that had never before been diagnosed. Doctors most often cited new and particularly malignant varieties of chickenpox. In the 1830s, there were fierce debates between doctors who advocated revaccination, considering these eruptions to be mitigated (or varioloid) smallpoxes, and those who, insisting that the preservative effect was permanent, instead saw these eruptive diseases as a variety of chickenpox.[69] Rayer's dermatological atlas thus made subtle distinctions between varioloid and chickenpox; between umbilicated, conoid, globular and vesicular chickenpox; and between cowpox, vaccinella and false cowpox. By depicting the different species of pustules on a neutral, timeless, rectangular plate, devoid of all commentary, he naturalised what were nonetheless highly controversial categories. After all, the nosological complexity of the field of pustular diseases directly resulted from the

---

68   Pierre François Olive Rayer, *Traité théorique et pratique des maladies de la peau*, vol. 1, Paris: Baillière, 1826, p. 421.

69   In London, Berlin and Milan, the problem of the duration of immunity was studied much earlier than it was in France. Doctors continued to publish on cases that seemed problematic, without facing censorship. In the 1800s the *Medical and Physical Journal* in London, the *Annali universali di medicina* in Milan and *Hufeland's Bibliothek der praktischen Heilkunde* in Berlin regularly reported cases of smallpox after vaccination. In 1807, the problem of recurrences seemed so serious that the House of Commons called for an inquiry. The strategy of clinical exoneration never won doctors' unvarying support. In 1818, during a smallpox epidemic in Scotland, many of the vaccinated developed rashes. In 1818, Dr Thomson, from Edinburgh, invented the 'varioloid' specifically to avoid having to worry about the clinical appearance of pustules: to be able to classify all rashes in the same category, and thus quantify the (in)effectiveness of the vaccine. See Charles Steinbrenner, *Traité sur la vaccine ou recherches historiques et critiques sur les résultats obtenus par les vaccinations et revaccinations*, Paris: Labé, 1846, pp. 31–180.

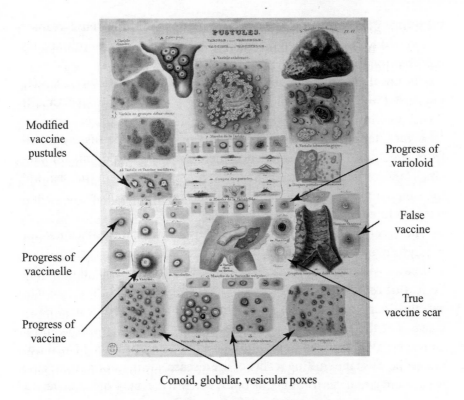

Modified vaccine pustules

Progress of varioloid

False vaccine

Progress of vaccinelle

True vaccine scar

Progress of vaccine

Conoid, globular, vesicular poxes

Pierre François Olive Rayer, 'Traité théorique et pratique des maladies de la peau', *Atlas*, 1835, © BNF

fierce struggles surrounding cowpox and its preservative attributes. In postulating the total visibility of the morbid, clinical dermatology produces the possibility of its own complication and allows for a honing of vaccine theory precisely by integrating its own dysfunctions.

The power of the image lies in its great capacity to circulate. Engravings were an essential factor in the global spread of vaccine theory. For instance, in Canton in 1803, the surgeon Alexander Pearson invited Chinese doctors to buy pus from the East India Company's stores and, to guide them in their practice, had a plate printed portraying pus as the 'genuine pustule'.[70] Cowpox was thus defined by depiction, for while treatises gathered dust on the shelves, engravings sold briskly. Husson's treatise, for example, found few takers: a bookseller in Reims did not want it 'for fear that he would be stuck with it'. On the other hand, he was extremely interested in the engravings.[71] The public had already made up its mind: the essence of cowpox could be put down on paper. Recognising

---

70   ARS, ms 94.
71   AAM V 57, Caqué to Husson, 8 germinal an IX (30 March 1801).

the primacy of the image, some doctors gave up publishing dense treatises and joined forces with the greatest engravers of the time to print plates representing the different periods of cowpox.[72]

The nosologies made up to protect vaccine from its critics were largely publicised well beyond medical schools. The prefects of the departments distributed instructions teaching the population about the existence of false vaccines and varioliform eruptions that had to be distinguished from smallpox.[73] Popularising such theories prepared parents for vaccine's failures to protect from smallpox, it convinced them that, behind the simplicity of the procedure, there stood expert diagnoses that could not be refuted.

The vaccinators pulled off a coup whose historical significance can hardly be overestimated. Never before had medicine used images to define a disease.[74] Anatomical atlases had long been used to chart the healthy body. But representations of pathologies were intended to document extraordinary cases rather than define types. The attempt to classify diseases at the end of the eighteenth century (by François Boissier de Sauvages, William Cullen or Philippe Pinel) had proceeded without images because the shifting reality of the disease – which varies from case to case and circulates through the body, where it can manifest in different places – prevented any pictorial definition of the disease. In 1800, no one was bound to accept the hypothesis that morbidity could be identified by a set of lines and patches of colour. Georges Cuvier emphasised the novelty of the graphic nosology promoted by the vaccinators and explained its limitations: 'As no person is precisely sick like any other, we can only give individual portraits of our infirmities, whereas, in regular beings, the individual represents the species.'[75] Dr Goetz mocked vaccinators for aping botanists: 'Tournefort and Jussieu have no better designed their learned system of plants, all in their own names at present . . . and the most brilliant flowerbed does not present a more delightful spectacle to the enchanted eye than the grey, white, red and charming pinkish

---

72   François Chaussier, professor of anatomy at the Paris School of Medicine, teamed up with one of the great Parisian engravers of the day, Louis-Pierre Baltard, to print large plates that were sold in great numbers. Husson commissioned his engravings from Jean Godefroy, a famous Parisian engraver, while Dr Fournier picked Antoine Cardon, the most renowned engraver in Brussels.

73   AN F8 113, interior minister to prefect of the Loire, 2 August 1810.

74   Barbara Stafford, *Body Criticism: Imaging the Unseen in Enlightenment Art and Medicine*, Cambridge, MA: MIT Press, 1991.

75   Georges Cuvier, *Rapport historique sur les progrès des sciences naturelles depuis 1789 et sur leur état actuel*, Paris: Imprimerie impériale, 1810, p. 343.

buds of vaccinia.'[76] What was at stake, in 1800, with the specific case of cowpox, was the possibility of defining a typical disease using graphics. The first dermatological atlases (by Robert Willan and Jean-Louis Alibert) appeared in 1805, followed by atlases of anatomical pathologies in the 1830s. The visual culture of nineteenth-century medicine was heir to the preferred nosological means of expressing the powers of vaccination.

6. Statistical utopia, informational dystopia

When, in 1804, the French government endorsed the theory of a perfectly benign vaccine, the public controversy ended immediately. The Interior Ministry decreed that any article on cowpox must be approved by the vaccine committee before publication.[77] The general press, which in 1802 and 1803 had published stories of accidents and relapses, was muzzled. The philanthropic committee became a central Vaccine Committee, placed under the authority of the Interior Ministry. Its members, the most influential Parisian doctors – including Michel-Augustin Thouret, director of the health school; Philippe Pinel, head doctor at Bicêtre; and Mongenot, head doctor at the children's hospice – were paid by the administration. Vaccine Committees were also set up in each *département* to correspond with the central committee in Paris. In 1804, one doctor called vaccinia 'the result of the perfection achieved by the government's science'.[78]

Above all, the administration enabled vaccinators to reorganise thoroughly the circulation of medical information. The aim of the 1804 circular introducing the vaccinia service was twofold: to make the benefits of the vaccine clear to the people; and, in return, to record the results of the vaccinations in order to corroborate their preservative virtue – not through medical expertise, but through the statistical tallying of the millions of children vaccinated throughout France. According to Chaptal, 'by keeping a record each year of the ever-decreasing number of those who have been attacked [by smallpox] and of the smaller proportion of its victims in the mortality lists, a general conviction will be established'.

---

76   François-Ignace Goetz, *De l'inutilité et des dangers de la vaccine*, p. 25.

77   AAM V6, letter from Fouché to Sauvo, editor of the *Moniteur universel*, 25 July 1809, obliging him to submit articles to the committee. Censorship must have existed before 1809, see AAM V1d4, Vaume, 'Mémoire confidentiel sur la vaccine', 22 July 1806.

78   AN F8 110, Vigaroux, Report to the vaccine committee on vaccination in the Hérault *département*.

Statistics were not the *sensorium* of the state, but provided a means of comparability, of highlighting the benefits of vaccination for the general public. On the one hand, they publicised the risk of smallpox: the prefects had a list of the names of those who had died of smallpox posted at the gates of each town hall, so that careless parents who had not had their children vaccinated would be exposed to public scapegoating.[79] On the other hand, by undertaking an exhaustive survey of all vaccines, it produced an accounting apparatus that would generate useful results.

Let's look more closely at how the numbers were functionally produced. The administration required vaccinators to fill in tables with six columns: the number of the person vaccinated, date, name, age, address and 'observations'. Having been accustomed to write up their cases in notebooks, vaccinators now had to record their observations in a narrow column. Either they left it completely blank, which is what most of them did, especially when they issued very long lists, or else they summarised the results of their work in a few lines.

The dangers of vaccinia can be seen in these notes, written in tiny characters. A vaccinator in the Hautes-Alpes noted with satisfaction: 'All the children vaccinated during 1806 had only a few small ulcers and scabs.' In 1806, a surgeon in Montauban, vaccinated only twelve people, but reported two miliary rashes that lasted twenty-one days; among the twenty people vaccinated by another health officer, there was one scrofulous tumour, one miliary rash and one scab.[80] In the Orne *département*, a vaccinator counted eight general rashes out of 276 vaccinated.[81] Dr Taulin, from Saint-Dizier, provided a table with 157 vaccinations; the 'Observations' column referred to eight accidents, two of which were fatal.[82] In 1812, in Nantes, Dr Valteau vaccinated thirty-two newborn babies at the hospice for foundlings; twenty-two died in the month following vaccination.[83] Dermatological waxes at the Saint-Louis hospital show the seriousness of post-vaccination dermatological affections, which were very common in the nineteenth century.

---

79   AN F8 113, interior minister to the prefect of the Loire, 2 August 1810: 'This authentic picture of parental negligence seems to me to be one of the most powerful means of forcing families that are still undecided to finally submit to a method that will keep their children with them and spare them the shame of a naming that covers them in blame.'

80   AAM V 54.

81   AAM V 63 d3, Orne, 'État des vaccinations de Gavelon', 1811.

82   'État des personnes vaccinées par J-B Taulin, résidant à Eurville, adjoint du comité de vaccine de St Dizié', 1808.

83   AN F8 113, Valteau to the prefect of Loire-Inférieure, 21 August 1813.

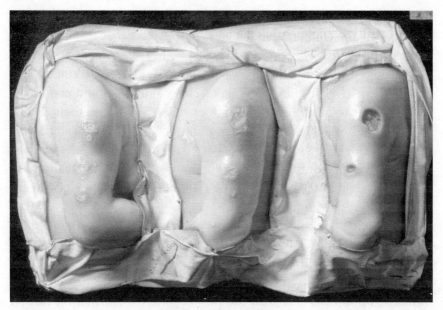

Ulcerative vaccines (*circa* 1880)[84]

Ignorance, like knowledge, had to be manufactured. In the case of vaccine risks, ignorance was created by a pyramid-shaped management of information, organised at several levels: town halls, departmental committees and the central committee, all of which operated as so many filters for bad news. Complications (ulcers, scabs, rashes that were sometimes dangerous) were reported in literary terms in the 'Observations' columns and were only rarely taken up by the upper echelons, which favoured numerical information for the purposes of quantification. Multiplying the stages in the transmission of information thus maximised the effects of the vaccinators' self-censorship. As the vaccine was supposed to be perfectly benign, the health officer or doctor who ran into an accident could fear that this would be chalked up to his own malpractice. For example, in 1820, in the Alps, the visit by two vaccinators produced hundreds of eruptive diseases. Out of 600 people vaccinated, 40 died. The *département*'s vaccinator accused the health officers of confusing or mixing vaccine pus with variola pus.[85]

Given that vaccination was poorly paid work, it was generally entrusted to simple health officers. They received meagre bonuses from

84   Dermatological waxes from the Saint-Louis hospital museum, Lariboisière, F. Widal hospital group, APHP, Paris.
85   AAM V25, letter from Dr Rabasse to the Prefect of the Hautes-Alpes *département*, 12 April 1820.

the departmental administration, based on the number of vaccinations they carried out. Unlike the inoculators of the 1760s, they had no interest in exposing their competitors' accidents. Unless they were particularly headstrong, then rather than risk being seen as an anti-vaccinator and expose themselves to the arrows of vaccine committees and prefects, it was far easier to keep their observations under wraps and their doubts to themselves. Statistics also had a moral function: the responsibility for keeping accidents quiet for the greater good of the nation, the work of refuting parents' complaints and exonerating vaccinia clinically were spread throughout the entire vaccine system. Each level had its share of accidents, scruples and indignities.

The vaccine committee, by organising and dominating the medical information network, controlled the circulation and the bad effects of accident narratives. The statistical system brought together in a single centre the scattered strands of a previously decentralised network with numerous points of access to the public sphere. Information flowed both upwards and downwards, with (self-)censorship at each point of transmission. Given its power to publish, censor, reward and punish, the central committee guided the judgement of vaccinators at the base of the pyramid. This management of information allowed the vaccine committee – which claimed to see better, more, indeed quantitatively – in fact to see only what it wanted.

To confirm this property of the vaccine system, which succeeded in isolating the committee in a world that was virtual and yet also convinc-ing, let us look at the healing effects of the vaccine. Unlike accidents, accounts of illnesses (scrofula, scabs, scabies, epilepsy) that disappeared during the operation were explicitly praised by the central committee, which mentioned them in its annual reports. In 1807, the committee invited doctors to seek out miraculous cures actively, and to send it reports.[86] The results were not long in coming: the vaccinators mirrored the cases that so interested the Parisian medical elite, even if they were not convinced of the causal link between the vaccine and the cure. But in departmental vaccine committees and then in the central committee, hundreds and thousands of similar cases built up, seeming to attribute a further role to this decidedly congenial virus.

Statistics produced an extremely powerful argument: in the main, vaccinations – that is, in the vast majority of cases reported in the mass of tables with their ever-blank columns of observations – were absolutely

---

86   Mémorial administratif no. 215, 3 March 1807.

safe. The small number of reports of accidents that managed to get past the successive hurdles of (self-)censorship and meticulous verification by the committee – meaning, in short, the few accidents or recurrences that remained inexplicable and forced their way onto the table of the central committee – were then weighed against the hundreds of thousands of problem-free vaccinations. Evidently, these accidents did not carry much weight, and in any case could not call into question the perfectly benign and perfectly protective vaccine.

Clinics and statistics were two ways of understanding the regular – two ways of defining the typical on different scales. They immunised vaccination from criticism by bringing into play the metaphysical distinction between essence and accident. They allowed vaccinators to formulate contradictory propositions, such as this conclusion to the 1807 report: 'It is not to be believed that the course of the new inoculation has not been troubled by a few storms, a few complications, or even a few accidents. But the sum of these is so small that no general conclusion against vaccinia can be drawn.' Vaccinia was no longer defined simply as the sum of its effects on a group of bodies (as had been proposed by the inoculators of the eighteenth century). Rather, it was defined by a subset of its effects which, in presenting greater consistency (that established by the clinic and statistics), constituted something like its essence.

From the 1820s onwards, vaccinia went through a long crisis. In the absence of booster vaccines (1840) and animal production of the vaccine lymph (1880), post-vaccine smallpox and contagions multiplied. Public mistrust persisted and even grew throughout the nineteenth century. This reticence owed to fear of bodily pollution, alteration of the blood and hereditary contamination. Contrary to the precepts of hygiene based on a strict separation between clean and dirty, cowpox seemed to dissolve the reassuring boundaries between one's own body and that of others, between the healthy and the pathological, and even between humans and animals.[87]

Despite this favourable climate, the French anti-vaccinatonists did not succeed in changing official medical doctrine. They did not try to collect and publish vaccine accidents, but rather sought to demonstrate the link between the vaccine and the *degeneration of* the population.[88] They did

87   Nadja Durbach, *Bodily Matters: The Anti-Vaccination Movement in England, 1853–1907*, Durham, NC: Duke University Press, 2005.

88   Anti-vaccinism was an early symptom of the medical, social and racial theme of degeneration. See Hector Carnot, *Petit traité de vaccinométrie*, Paris: Moquet, 1849, 1857; Henri Verdé-Delisle, *De La Dégénérescence physique et morale de l'espèce humaine*

not analyse the construction of the statistics produced by the committee but settled for correlating them with other data sets: average heights for each birth year, the birth rate, disease statistics, and so on. By trying to show too much, the French anti-vaccinators failed to convince anyone: they created a lot of confusion and little mobilisation.

The first awareness of contamination came from doctors who simply wanted to make money from vaccinia. To do this, they needed to create a market outside the free vaccinia offered by the committee, to which end they promoted their own purer, safer lymph. The first experiments with bovine vaccines were connected to the demand coming from notables. As early as 1803, the central committee used cows to satisfy some bourgeois who refused the humanised vaccine.[89] At the end of the 1820s, Dr James founded the Société nationale de vaccine to propagate the heifer vaccine. According to the Académie de médecine, by promoting the animal vaccine, Dr James was discrediting vaccinia in the eyes of all. Because the production of the animal vaccine was difficult and costly, French vaccinators consistently denied that the humanised vaccine could bring contamination.

The slowness of the process that led to the official acceptance of the risks well illustrates the resilience of the vaccine system. It was not until 1864 that the Académie de médecine accepted the possibility of syphilitic transmission. And it was not the vaccinators who, debating the risk of contamination, came to study syphilis; it was the students of syphilis, in the midst of a dispute over the possibility of this disease being transmitted by blood, who turned their attentions to vaccination, making it into a theme of their own polemics.[90]

Vaccination led to a significant divide between doctors and the public. For in the very moment that this practice caused hundreds of deaths and thousands of accidents, as parents complained to vaccinators who countered them with statistics and rebuttals put forward by clinical expertise, and as the official doctrine was repeated on the one side and rejected on

---

déterminée par le vaccin, Paris: Charpentier, 1855; Armand Bayard, Influence de la vaccine sur la population, Paris: Masson, 1855. The great theoretician of degeneration, Bénédict Morel, took care to distinguish his theories from those of the anti-vaccinationists (Traité des dégénérescences physiques, intellectuelles et morales de l'espèce humaine, Paris: Baillière, 1857, p. 560).

89   Rapport du comité central de la vaccine, p. 379. See also Le Journal du soir no. 1025, 25 ventôse an IX (16 March 1801).

90   See Jean-Baptiste Fressoz, 'Le vaccin et ses simulacres: instaurer un être pour gérer une population, 1800–1865', Tracés. Revue de sciences humaines, 21, 2011, pp. 77–108.

the other, in doctors' eyes the public itself was transformed. From being a locus of authority able to pass judgement and produce information, a rational body that had to be convinced by hard numbers, the public became an inert mass to be subjugated by medical and administrative authority, making its own incompetence explicit. A new genre appeared in medical literature: dissertations on 'popular errors and prejudices'.[91]

It would be easy to ridicule the doctors who vilified the so-called 'popular prejudices' that turned out to be the precautions that vaccinators of subsequent generations would have to heed. But the history of cowpox has other meanings. The committee played the role assigned to it, that is, as a screen to hold back bad news and cast a comforting shadow play. In essence, it allowed politicians and individuals to offload the moral weight of the decision. Unlike smallpox inoculation, which was embroiled in never-ending ethical discussions on God, nature and politics, and unlike risk with its unsolvable moral and psychological dilemmas, expertise on vaccination could maintain the idea of a perfectly benign practice, a simple technique without moral consequences. Far more effectively than risk, expertise produced disinhibition: by administering proof and also ignorance, it was the condition for the general acceptance of the vaccine and, once again, for changing the scale at which humanity itself existed.

---

91   Anthelme Richerand, *Des Erreurs populaires relatives à la médecine*, Paris: Caille, 1812; Pierre Adolphe Piorry, *Dissertation sur le danger de la lecture des livres de médecine par les gens du monde*, Paris: Didot, 1816. One subgenre dealt explicitly with vaccinia: Jean-Baptiste Dugat, *Erreurs et préjugés populaires sur la vaccin et la petite vérole*, Avignon: Guichard, 1823; Jean-Étienne Thorel, *Dissertation sur les préjugés populaires qui s'opposent à l'adoption générale de la vaccin*, Strasbourg, 1823; Dominique Latour, *Réfutations de quelques préjugés sur la vaccin*, Toulouse: Chez l'auteur, 1823; Cortade, *Manuel de vaccine dédié aux officiers de santé et habitants des campagnes du département du Gers*, 1819.

3

The Ancien Régime and the Environment

Modern environmental destruction played out not in a world where nature was unimportant, but in one whose dominant theories saw the environment as fundamental to producing human beings themselves. To understand this paradox, we need to get away from our dualisms of innate vs acquired and body vs environment, and instead think within an epistemic space that has now disappeared. This means thinking in terms of a theory of climates in which human action, the environment and bodies were intertwined.

The notion of climate that Montesquieu popularised in his *De L'esprit des lois* does not do justice to the richness of this concept. In the eighteenth century, climates were not just large spaces defined by their latitudinal situation, little controlled by mankind. Rather, this notion spanned a great deal more: it included the 'airs, waters, and places' of Hippocratic inspiration and, more generally, all the *circumfusas* or 'surroundings' that influenced health and moulded bodies.[1] In this respect, a seminal work of climate determinism was Jean-Baptiste Dubos's *Les Réflexions critiques sur la poésie et sur la peinture* (1719): here, the climates that determine peoples' physical and intellectual qualities are produced by local environments

---

1   Clarence J. Glacken, *Traces on the Rhodian Shore: Nature and Culture in Western Thought from Ancient Times to the End of the Eighteenth Century*, Berkeley: University of California Press, 1967; James Riley, *The Eighteenth-Century Campaign to Avoid Disease*, London: Macmillan, 1987. On *circumfusa*, see Louis Macquart, entry 'Climat' in *Encyclopédie méthodique. Médecine*, Paris: Panckoucke, 1792, vol. 4, p. 878. The term *circumfusa was* popularised in the eighteenth century by the physician Jean-Noël Hallé and his division of hygiene into *circumfusa, ingesta, excreta, percepta* and *gesta*. See Gérard Jorland, *Une Société à soigner, hygiène et salubrité publiques en France au XIXe siècle*, Paris: Gallimard, 2010, pp. 42–5.

transformed by human action. In his efforts to explain why Romans had degenerated since ancient times, the author cited the destruction of the sewers (*cloaca maxima*) by the barbarians and the spread of alum mines, which had affected the air in the city.[2] Similarly, in 1731, in his influential *Essay Concerning the Effects of Air on the Human Body*, the physician John Arbuthnot described air as a mixture of natural and artificial exhalations that determine human health.[3]

Human societies were seen as evolving in relation to atmospheric surroundings which they themselves fashioned. The climate made the sum total of all possible environmental transformations; technical activity echoed through the climate, which in turn altered the constitutions of humans themselves.

In the eighteenth century, the environment was first and foremost an issue of biopolitics: since the *circumfusa* had a decisive influence on health, governments could use them to influence the number and strength of their subjects. For example, the treatise by Jean-Baptiste Moheau and Antoine Montyon on demography and political economy, which taught the sovereign how he could increase the population (via sound laws on trade, taxes and inheritance), concluded with a programme that was simultaneously environmental, populationist and anthropotechnical: 'It is not only through . . . useful institutions . . . that Kings can favour population; *the whole physical order* seems to be still in their hands.' And, since 'a different climate forms a new species', a sovereign who properly managed the *circumfusa* of his realm could take control over the health, size and even the shape of his population.[4] Abbé Richard, author of a ten-volume *Natural History of Air*, explained that studying this subject was 'not mere speculation' but 'useful for the great art of governing men'.[5] In 1776, the French monarchy founded the Royal Society of Medicine to guide its medico-environmental policy. At its behest, doctors drew up 'medical topographies' which described in minute detail the environmental surroundings of various locations and their influence on the inhabitants' health.[6] In the eighteenth century,

---

2   Jean-Baptiste Dubos, *Réflexions critiques sur la poésie et sur la peinture*, 1719, Utrecht: Étienne Neaulme, 1732, vol. 2, pp. 152–7.

3   John Arbuthnot, *An Essay Concerning the Effects of Air on Human Bodies*, London: Tonson, 1731, p. 10.

4   Jean-Baptiste Moheau and Antoine Montyon, *Recherches et considérations sur la population de France*, Paris: Moutard, 1778, vol. 2, p. 156.

5   Jerôme Richard, *Histoire naturelle de l'air et des météores*, Paris: Saillant, 1770, vol. 1, p. 2.

6   Jean Meyer, 'L'enquête de l'Académie de médecine sur les épidémies, 1774–1794', *Annales ESC*, vol. 21, no. 4, 1966, pp. 729–49.

the many different forms of knowledge about air, its hygienic condition and its composition (air chemistry, pneumatics, eudiometry, meteorology and medical topography) came together in this biopolitics of atmospheres.[7]

The power of the *circumfusa* was also a source of concern. For as humanity altered its surroundings, it also risked changing itself. Seemingly benign environmental transformations could have terrible consequences. According to Richard, an epidemic in the Dutch Moluccas was caused by the destruction of clove trees, whose aromatic particles had hitherto counteracted the corruption of the air by a volcanic release.[8] New diseases could be products of human action: syphilis was purported to have originated in the mines of Santo Domingo, where the Spaniards subjected the *indios* to such appalling working conditions that it changed even their physical constitution.[9] The smoke from the workshops gave rise to similar concerns among the urban bourgeoisie: in the eighteenth century, cities were the insalubrious locations *par excellence*, in the same vein as swamps, prisons and ships.[10]

As industrialisation advanced, its attendant pollution and massive use of natural resources radically transformed its surroundings. All this took place within the theoretical framework of climate medicine. So the problem posed to historians is not a matter of understanding how so-called environmental awareness ultimately emerged. Quite the opposite: the question is how to understand the schizophrenic nature of industrial modernity, which continued to conceive man as a product of his surroundings, even as it allowed him to alter and destroy them.

In France, the emergence of chemical capitalism was a decisive factor in the industrialisation of the environment. In the 1800s, the chemical industries provided a historic meeting point between massive pollution, new production methods, a considerable volume of capital and the scientific and administrative elite that had emerged from the Revolution. This combination of innovation, profit and power made it possible to transform environmental regulations, as required by the development of manufacturing capitalism. Seeking to shed light on the industrial disinhibition that

---

7   Simon Schaffer, 'Measuring virtue. Eudiometry, enlightenment and pneumatic medicine', in Andrew Cunningham and Roger French (eds), *The Medical Enlightenment of the Eighteenth Century*, Cambridge: Cambridge University Press, 1990, pp. 281–318.

8   Richard, *Histoire naturelle de l'air et des météores*, vol. 2, p. 412.

9   Jean-Bernard Bossu, *Nouveaux voyages aux Indes occidentales*, Amsterdam: Changuion, 1769.

10   Alain Corbin, *Le Miasme et la Jonquille*, Paris: Aubier, 1982; Sabine Barles, *La Ville délétère. Médecins et ingénieurs dans l'espace urbain*, Seyssel: Champ Vallon, 1999.

marked the 1800s, this chapter surveys the Ancien Régime regulation of artisanal environments around the mid-eighteenth century.

1. Policing the arts and the air

At that time, the *police* force was much more than just the contemporary institution for maintaining order; rather, under the Ancien Régime, it fully deserved its name. For the regulations it promulgated, the surveillance it enacted, and the reprimands it imposed shaped the urban context and modes of inhabiting the *polis*.[11] It had a vast remit: maintaining order, keeping up supplies to towns and cities, inspecting the safety of transport and buildings, preventing fires, monitoring markets and checking foodstuffs, ensuring the cleanliness of the streets, and so on. What Michel Foucault once called a police that 'includes everything', the Ancien Régime police force was, more precisely, a policing of all of man's *surroundings*: and it was the *circumfusa* and their importance for public health that justified its hold over the city.[12] According to Prost de Royer, lieutenant-general of the Lyon police, his force could limit the height of buildings and define the width of streets and courtyards because 'the preservation of public health, through the salubrity of the air, unquestionably gives it this right'.[13] The stakes were high: the health, numbers and even shape of the population depended on the proper management of urban environments. The Rouen doctor Lepecq de la Clôture attributed the disastrous health of Parisians to the negligence of the police, who 'foment or perpetuate contagious diseases': 'The inhabitants of Paris *have been made* the weakest and unhealthiest people on earth.'[14] The Parisian police *commissaire* Delamare explained that the purpose of his institution went beyond the good health of the individual to include 'the integrity and perfect conformation of his limbs'.[15]

---

11   See the masterly study of the Parisian example by Thomas Le Roux, *Le Laboratoire des pollutions industrielles. Paris, 1770–1830*, Paris: Albin Michel, 2011; Jérôme Fromageau, *La Police de la pollution à Paris de 1666 à 1789*, PhD in Law, Paris II, 1989.

12   Michel Foucault, '*Omnes et singulatim*: towards a criticism of political reason', in *The Tanner Lectures on Human Values*, Cambridge: Cambridge University Press, 1981, vol. 2, pp. 223–54.

13   Antoine-François Prost de Royer, *Dictionnaire de jurisprudence et des arrêts*, Lyon: Roche, 1783, vol. 3, p. 746.

14   Louis Lepecq de la Clôture, *Collection d'observations sur les maladies et constitutions épidémiques*, Rouen: Imprimerie privilégiée, 1778, p. 26.

15   Nicolas Delamare, *Traité de la police*, Paris: Jean-Pierre Cot, 1705, p. 533.

The police also justified its power over the city in terms of its ability to manage the risks created by concentrations of buildings, people and activities. The concept of *imminent danger* was fundamental, as it called for surveillance in all places and moments, instant governance, and a capacity for anticipation and prevention that only the police could claim to possess. The royal decree of 1729 entrusting the Paris police with the safety of public thoroughfares strengthened surveillance and formalised streamlined procedures. Police governance had to be swift:

I. The commissioners will pay particular attention to each of their districts in order to be informed of the houses and buildings where there is some danger.

II. As soon as they are informed of this, they will go to the place and draw up a report of what they have noticed there, and which could be contrary to public safety.

III. They will have the owners summoned without delay at the request of our Châtelet prosecutor on the first day of the hearing of the police of our Châtelet de Paris.[16]

In Paris, forty-eight police *commissaires* in twenty neighbourhoods divided up the entire urban space.[17] A large part of their work concerned cleanliness and compliance with urban regulations (sweeping in front of the house, handing over one's rubbish to the contractors responsible for taking it away, not hanging anything from one's windows, locking the door at night, and so on). Each day, the police *commissaire* had to do his rounds through his neighbourhood of operations (where he also had to live) wearing a distinctive gown and accompanied by a bailiff.[18] Despite this judicial pomp, it appears that there was no matter too small or too trivial for the police. Police *commissaire* Lemaire thus notes:

---

16   Nicolas Delamare, *Continuation du traité de la police*, Paris: Hérissant, 1738, vol. 4, p. 127.

17   Alan Williams, *The Police of Paris: 1718–1789*, Baton Rouge: Louisiana State University, 1979; Paolo Piasenza, 'Juges, lieutenants de police et bourgeois à Paris aux XVIIe et XVIIIe siècles', *Annales*, vol. 45, no. 5, 1990, pp. 1189–215; Vincent Milliot, 'Qu'est-ce qu'une police éclairée? La police amélioratrice selon Jean-Charles Pierre Lenoir, lieutenant général à Paris (1775–1785)', *Dix-Huitième Siècle*, 37, 2005, pp. 117–30.

18   Jean-Baptiste Lemaire, 'Mémoire sur l'administration de la police en France (1771)', *Mémoires de la Société de l'histoire de Paris et de l'Île-de-France*, 1878, vol. 5, pp. 29, 58.

Police of 17 July 1768, rue Mouffetard entering via rue Bordet: a female individual shelling peas and leaving her shells on the pavement in front of the shop on the right where she lives, first summons; rue d'Orléans entering via rue Mouffetard: a considerable pile of earth and garden refuse obstructing the public thoroughfare, second summons; quai de l'Horloge du Palais, two flower pots on the window on the first floor above the mezzanine and the shop where is written 'Sauvage bijoutier', third summons, etc.[19]

The repeated injunctions, reminders of regulations, summons to Le Châtelet and the issuing of fines were meant to make the city clean and secure. The police's ongoing work shaped individual behaviour and civilised urban mores by bringing them into line with the common good. For Lemaire, 'the police is . . . the science of governing men and doing them good, the way of making them, as much as possible, what they should be for the general interest of society.'[20]

The presence of artisans in the heart of the city was regulated by this policing of urban micro-risks: the workshops, their fumes and vapours, their overly intense fireplaces and their run-off were subject to a police procedure similar to that regarding 'imminent perils'. This regulation by police surveillance was well suited to the artisan world: since their productive activity was conceived as a set of gestures and tricks of the trade, the nuisances it caused could not be regulated either by the enactment of a technical standard that would pre-empt the said nuisances, or by workshop rules that would prescribe the correct operating procedures. These tacit knowledges, secrets of production and different ways of doing things in each workshop explain the police's agnosticism over their systems: the productive infrastructure was not an object that the police could control, amend or govern, and so it was not there that it looked to resolve the environmental problem. Conversely, monitoring the effects of production was central: environmental regulation was produced through a continual work of policing trades, rather than via any specific definition of technical forms or production processes.

One example is the case of the workshops distilling 'vitriol oil' (*eauforte* or nitric acid), which is well documented thanks to the precise description given by the pharmacist Jacques-François Demachy. Because chemistry was a bottleneck for the textile industry booming in the 1760s,

---

19   AN Y9 471 B, Commissaire Lemaire, Rétif huissier, police 16 July–4 August 1768.
20   Lemaire, 'Mémoire sur l'administration de la police en France (1771)', p. 28.

Jacques-François Demachy, Description of the art of the vitriol distiller, 1773[21]

and because know-how was scarce, Demachy paid scrupulous attention
to the work of the artisan, his skills, and the sensations (the aroma of the
oil, the odour of the fumes) that went into making a good product. While
the plates show the worker as being cramped and solitary, a little lost in a
gigantic workshop, in the text he is constantly in action. His movements
are not subject to analytical rationality: there are no fixed procedures,
only long shifts during which the worker has to take various initiatives.
The distillation of vitriols to produce vitriol oil took around twelve hours.
The worker had to constantly plug the cracks in the stoneware retorts
and sniff the vapours to decide whether to let out the ones not made up
of vitriol oil. Wastage was the norm and leakages were prevented only
thanks to workers' constant vigilance: 'Nothing is more common in
distillation than to see the dome split . . . which gives the worker a lot of
work to plug these cracks, as they let out the red vapours.'[22] For Demachy,
this technique was essentially hazardous. The materials used (sandstone,
clay, loam, dung) were porous and easily crumbled, the retorts chafed,
burst and cracked – and these inconveniences were the *normal* state of
the technique. In a way, the distinction between artisan skills and tools

---

21   Neuchâtel, Société typographique, 1780.

22   Jacques-François Demachy, *Description de l'art du distillateur d'eau-forte*, Paris,
1773, pp. 19–21.

was irrelevant: the technique, which was almost impossible to pin down, was constantly produced and reproduced by the worker who plugged the cracks, sealed the leaks, smoothed the retorts and repaired the furnaces.[23]

In Paris in the 1750s and 1760s, six vitriolic acid workshops were scattered around the Porte Saint-Denis and Place Maubert. They were subject to the daily supervision of the neighbourhood commissioners, who checked during their rounds that the smoke was not excessive. Although there were no police regulations governing vitriol oil distillers, the commissioners regularly prosecuted them, invoking the general principles of public health.[24] The policing of the air and imminent perils opened the way for preventive governance: they justified punishing an action (or lack of action) that had caused no accident, did not stem from any desire to cause harm and contradicted no formalised norm.

The aim of police proceedings against an establishment was to have it moved outside the city walls. The commissioners did not ask for an expert opinion on the work processes, nor did they contemplate the possibility of improving them. Des Essarts, the author of an important police dictionary, considered the emissions from *eau-forte* workshops to be inevitable and, quite logically, argued for them to be banned from the city.[25] In 1768, the Paris *parlement* ordered the distiller Jacquet in the Maubert district to destroy his furnaces.[26] In 1786, the *intendant* of Provence authorised a vitriol oil manufactory on condition that it was established 'in the suburbs and isolated areas'.[27] Either the improvement of processes did not appear to be a solution to environmental conflicts, or the technique was beyond the control of the police, who fell back on the traditional solution of relocation. In Paris, the same logic of transferring the nuisance outside city walls prevailed for all trades based on the processing of animal products: gut-string-makers, tripe butchers,

---

23 Historians have shown the political dimension of the *Encyclopédie*, which did battle against trade secrets in order to expropriate the artisans' expertise; see Georges Friedman, 'L'*Encyclopédie* et le travail humain', *Annales, histoire, sciences sociales*, 8, 1953, pp. 53–68; William H. Sewell Jr, *Work and Revolution in France: The Language of Labor from the Old Regime to 1848*, Cambridge: Cambridge University Press, 1980. But as far as the chemical trades were concerned, this project remained incomplete: tacit artisanal knowledge eluded scholarly endeavours, which had neither the political nor the cognitive means to fulfil their ambition.

24 Delacoste in 1750 (AN Y9 533), Jacquet in 1768 (AN Y9 471 B) and Charlard in 1773 (AN F12 879).

25 Toussaint Le Moyne Des Essarts, *Dictionnaire universel de la police*, Paris: Moutard, 1786, vol. 6, pp. 1–2.

26 AN F12 879.

27 AN F12 1507, letter from the *intendant* to Aix, 1786.

tawers, curriers, glue-makers, starch-makers, brewers and tanners were excluded from the city as complaints arose and as the balance of power shifted between guilds, police and citizens. The environmental problem met with exclusively spatial solutions: the aim was to reduce as far as possible the conflicts among neighbours caused by the artisan, while also ensuring supplies for the city.[28]

In short, police regulation was conducted by the continuous surveillance of productive activity, by the recording of residents' complaints, and by repeated injunctions, threats, fines and bans, rather than by the enactment of technical standards.

2. The rules of the trade

Safety standards were the responsibility of guilds. While these latter had only a limited role to play in regulating the nuisances resulting from artisan production, they did play a fundamental safety role in that they embodied the rules of the trade.[29] This was an essential concept in the world of guilds: these rules separated tradesmen from handymen; they drew the line between order and disorder; they presupposed an intellect-in-action, 'a method for executing a thing well according to certain rules'; and finally, because they defined standards of quality and solidity, they helped to ensure safety in the technical worlds of the Ancien Régime.[30]

Generally speaking, the police force did not set safety standards. It was not that it was incapable of doing so (in 1672, an ordinance codified the proper construction of chimneys in precise terms, also using graphics). Rather, it preferred to rely on the guilds' own best practice. For example, when the Paris police force issued an ordinance on messenger services in 1779, it did not go into any technical details (the text simply required 'well-conditioned carriages') and referred to the statutes of the community of coachbuilders for further details.[31] Similarly, the safety of

---

28   Le Roux, *Le Laboratoire des pollutions industrielles*, pp. 46–68.

29   Although some statutes did incorporate the locational constraints imposed by the police and *parlements*, in general they simply duplicated the police regulations. For example, in 1744, in order to move the starch producers from the Seine to the Bièvre, the *parlement* of Paris added an article to the statutes requiring them to set up outside the capital. See *Encyclopédie méthodique. Arts et métiers mécaniques*, Paris: Panckoucke, 1782, vol. 1, p. 20. On the guild's 'internal-policing' role, see Paolo Napoli, *La Naissance de la police moderne. Pouvoir, normes, société*, Paris: La Découverte, 2003, ch. 3.

30   Entry 'Art' in *Le Grand Vocabulaire françois*, Paris: Panckoucke, 1768, vol. 3, p. 115.

31   Joseph-Nicolas Guyot, *Répertoire universel et raisonné de jurisprudence civile et criminelle*, Paris: Panckoucke, 1783, vol. 64, p. 187.

Parisian buildings relied on the masons' guild and, more specifically, its judicial body: the Chambre des bâtiments. This institution, which was responsible for monitoring building sites and settling disputes between contractors and master builders, produced standards through case law: sentences condemning faulty workmanship circulated throughout the Parisian construction community (for example, by being posted on building sites) and thereby indirectly defined good practice in this industry.[32] The safety of the workshops was also based on a guild standard that was built up successively through reports of faulty workmanship and not on formalised technical or architectural knowledges.

The guild system also played a role, albeit one that was difficult to grasp, in providing security within the workshop. In the eighteenth century, workmen's guilds financed relief funds for sick or injured workers and were thus well placed to understand the specific ailments of their trades. Take, for example, the conflict that divided the millinery industry in Marseille over the use of mercury to felt rabbit hair.[33] In 1774, the master milliners of Marseilles officially enshrined this technique, which until then had only been tolerated. The hatters' boys (the workers) appealed to the aldermen, who organised a medical consultation: four doctors publicly examined the workers and the Hôtel-Dieu hospital issued certificates of illness. Their report incriminated the master hatters of Marseilles, who used a mixture of vitriol oil and mercury that was much more concentrated than that used by their Parisian colleagues. In 1776, the *parlement* of Aix settled the dispute, authorising the use of mercury but forcing the masters to reduce the proportion. It also asked the community's sworn wardens and the police to check the mixture through frequent workshop inspections.[34] Because they were organised into a powerful *generalité* with nearly 600 members, the hat-makers of Marseilles were able to reach a compromise: the master hatters had to disclose and modify the preparation of the felting mixture. The guild system played an ambivalent role, here: the community of masters long proved reluctant to take workers' complaints on board. Yet once the decision had been taken, it undoubtedly

---

32   Robert Carvais, *La Chambre royale des bâtiments. Juridiction professionelle et droit de la construction à Paris sous l'Ancien Régime*, PhD in Law, Paris II, 2001.

33   Michael Sonenscher, *The Hatters of Eighteenth-Century France*, Berkeley: University of California Press, 1987, pp. 106–17; Abbé Tenon, 'Mémoire sur les causes de quelques maladies qui affectent les chapeliers', *Histoires et mémoires de l'Académie des sciences*, 7, 1806–7, pp. 98–116 (written in 1756).

34   Magnan, 'Mémoire sur les accidents auxquels sont exposés les garçons chapeliers de la ville de Marseille et sur les moyens de les prévenir', *Observations sur la physique*, 7, 1776, p. 148.

became an ideal institution for imposing compliance with a bothersome reform on each of its members as it was now a matter of ensuring fair competition (since higher concentration of mercury sped up production).

In the chemical trades, which were not very capital-intensive and heavily relied on know-how, the worker held considerable power. Following the principles of *louage d'ouvrage* stipulated by the civil law of the Ancien Régime – a sale of a specific service, rather than employment for a specific period – he committed to obtain some result, without any implication of obedience to the boss's command. The worker simply committed to deliver a product by a given date, and he remained responsible for the methods used. The tradesman's freedom in his work legally distinguished him from the domestic servant.[35] In the case of the chemical trades, workers largely set the methods of production, and thus the risks which they agreed to take. The working life of Pierre Mathieu – a craftsman of Armenian origin who held the 'secret of Cyprus blue vitriol' – demonstrates the power conferred by know-how. In 1762, his employer, Étienne de Lyon, was accused of murder and sent to the galleys.[36] Two entrepreneurs then fought over his services. The first settled his debts and tried to obtain an annuity to his benefit; the second, for want of his help, bought the secret of vitriol from Étienne de Lyon's wife, but was unable to obtain good-quality vitriol. The secret, without the worker who knew how to use it, remained worthless. For entrepreneurs, defection or betrayal by the worker who held the secret represented a considerable economic risk. Hence the stratagems aimed at securing the loyalty of this valuable and intractable workforce: getting them to sign loyalty contracts, obtaining various privileges (like exemptions from serving in the militia) or even annuities to their benefit.[37]

This also explains the growing concern over illnesses among tradesmen. In 1760, a doctor in Aachen congratulated himself on having treated a workman because 'his withdrawal would have been detrimental to our whole city'. For Tenon, the illnesses among tradesmen harmed 'trade and public fortunes because . . . they sometimes strip us of processes that are difficult to find anew'.[38]

---

35   On this crucial distinction between the hiring of work and the hiring of services, see Michael Sonenscher, *Work and Wages: Natural Law, Politics and the Eighteenth-Century French Trades*, Cambridge: Cambridge University Press, 1989, pp. 69–72.

36   AN F12 1506 and Liliane Hilaire-Pérez, 'Cultures techniques et pratiques de l'échange, entre Lyon et le Levant: inventions et réseaux au XVIIIe siècle', *Revue d'histoire moderne et contemporaine*, vol. 49, no. 1, 2002, pp. 89–114.

37   AN F12 1506, Holker to Trudaine de Montigny, 31 March 1768.

38   Boucher, 'Observation sur une maladie singulière des artisans', *Journal de*

3. Surveillance by neighbours: health and easements

Alongside the police and the guilds, the burghers also played an essential role: the surveillance of craftsmen relied above all on the vigilance of their neighbours. At the building, street or neighbourhood level, groups took form and mobilised against a given workshop. The police's work was based on these complaints. In 1750, the commissioner of the Faubourg Saint-Martin received Lady of Châtillon, who owned a building in the district. She complained about a glue manufacturer who was inconveniencing 'all the neighbours, both in the head and in the stomach'.[39] The noblewoman 'required' a visit from the *commissaire*, whose report was produced at the request of the complainant, who led him through the neighbourhood to observe the damage being done.

Following a complaint and a report, the crown prosecutor could launch judicial action. The *commissaire* then carried out an *information* (inquiry, or consultation), part of the criminal procedure. Unlike civil proceedings, where each of the parties had to 'provide their proofs' to a court clerk (and pay for the privilege), in a case having to do with industrial nuisances the inquiry was conducted by a *commissaire*, who would summon witnesses. For the police, the neighbours of a workshop were not defending their particular interests but testifying to facts concerning order and the public good. In this period, the police force occupied a different place in society to the prefectural administrations and civil courts of the nineteenth century: it saw itself not as an arbiter of conflicting individual interests, but as a protector of the common good, of which the neighbours were in the forefront.

Take the information on a vitriol oil workshop, drawn up in 1750 by a Paris police *commissaire*. The document lists twenty-four depositions from witnesses of widely varying statuses: master ribbon makers, workmen, bourgeois citizens, caterers and an advisor to the king. The lexicon used to describe pollution was highly standardised. Whatever the status of the witness, whatever the workshop, the same themes always came up: abundant fumes, foul odours and the illnesses they caused. Since the main purpose of these depositions was to trigger legal proceedings, they cannot be understood without reference to the way in which eighteenth-century jurisprudence identified the various phenomena that we would classify as pollution.

---

médecine, 12, 1760, p. 21; Tenon, 'Mémoire sur les causes de quelques maladies', p. 90.
39   AN Y9 533.

The question of public health came first: the witnesses systematically mentioned their own illnesses and those of their neighbours. They emphasised the temporal or spatial correlations between the workshop and their ailments, in accordance with the prevalent environmental aetiologies of the time. The second argument concerned the intrusion of smoke into the private space, which has the advantage of falling into the old and well-established categories of easement law. For eighteenth-century jurists, the smoke from workshops was an unnatural easement (unlike a watercourse crossing an estate, for example) illegally imposed on the owner.[40] Although easements could be bought and passed on, Ancien Régime jurisprudence was careful to specify that nuisances caused by craftsmen were an exception to these arrangements between private individuals.[41]

We should remember this point: according to police procedure, it was the neighbours who proved the existence of pollution. According to the canons of judicial proof under the Ancien Régime, based on the number and quality of testimonies, the numerous and concordant depositions of the burghers bore great evidential weight. The neighbourhood police commissioners assisted them in this task. Without wishing to prejudge their honesty, it is nevertheless likely that their interests overlapped with those of propertied rentiers and not those of the artisans. Indeed, the venality of the office of *commissaire* encouraged the formation of bourgeois dynasties of policemen, rooted in districts threatened by the artisans' advance.[42]

4. The power of notables and the *commodo incommodo* consultation

As far as police forces and *parlements* were concerned, in principle it was up to local notables to decide on how spaces should be used. The *commodo incommodo* information procedure was evidence of this power. In the eighteenth century, this prior consultation was commonplace. Any institution (including the king, police force, *parlements*, aldermen,

---

40   Entry 'Voisinage', in *Encyclopédie méthodique. Jurisprudence*, Paris: Panckoucke, 1789, vol. 8, p. 181; entry 'Servitude', in Jean-Baptiste Robinet (ed.), *Dictionnaire universel des sciences morale, économique, politique*, Paris: Libraires associés, 1783, vol. 28, p. 185.

41   Antoine Desgodets, *Les Loix des bâtimens* [sic], *suivant la coutume de Paris*, 1748, Paris: Libraires associés, 1787, p. 55.

42   Vincent Milliot, 'Saisir l'espace urbain: mobilité des commissaires et contrôle des quartiers de police à Paris au XVIIIe siècle', *Revue d'histoire moderne et contemporaine*, vol. 50, no. 1, 2003, pp. 54–80.

and water and forestry authorities) could order an inquiry in order to 'enlighten its religion'. The distinctive feature of the procedure was that it was pre-emptive. Its stated aim was to identify the particular interests that could be affected by a decision and to weigh them up against the expected public and private benefits. The areas in which usually applied were at the intersection of the public interest and private property: typically expropriations, but also the organisation of fairs, street alignments, the development of waterways, the establishment of game reserves, the draining of marshes, hospital and ecclesiastical properties and parish assets.[43] *Commodo incommodo* was a common feature of urban management. Municipalities began to take charge of the shape of the city in the name of efficiency, health and the public interest used this consultation to 'enlighten their religion', prevent complaints and circumvent any opposition from burghers or guilds by invoking the general interest of the city.

In Paris, throughout the eighteenth century, a preliminary consultation of the inhabitants was required for a growing number of harmful establishments: tallow foundries, tanners, starch manufacturers, glue factories, gut factories, and so on. This was no mere formality: the surveys that have been preserved contain a detailed site plan and letters of support or refusal from neighbours.[44] To authorise a simple butcher's stall, the commissioner had to consult at least 'twelve notable burghers' and the syndics of the community of butchers.[45] Factories applying for royal privileges were also subject to the *commodo incommodo* procedure. Indeed, before registering privileges, the *parlements* had to carry out an inquiry to check whether the derogation from general law might be likely to harm private individuals. Paradoxically, it was because the crown encouraged and protected a manufactory that it could be pulled into the dock by the *parlement*.[46] In Rouen, there were two stages to the consultation on the first major chemical factories between 1760 and 1770. The first stage was very open: all the inhabitants of Rouen were invited by poster to submit their opinions to the *parlement*'s clerk. The second stage consisted of

---

43   Delamare, *Continuation du Traité de la police*, pp. 266, 329, 668, 863; Jean-Baptiste Denisart, *Collection de décisions nouvelles et de notions relatives à la jurisprudence*, Paris: Desaint, 1775, vol. 2, pp. 337, 341, 573; Joseph-Nicolas Guyot, *Répertoire universel et raisonné de jurisprudence*, Paris: Visse, 1784, vol. 8, p. 129; M. Chailland, *Dictionnaire raisonné des eaux et forêts*, Paris: Ganeau, 1769, vol. 2, p. 309.

44   AN Y9 504.

45   Entry 'Étal', in *Encyclopédie méthodique. Jurisprudence*, Paris: Panckoucke, 1791, vol. 10, p. 157.

46   Pierre-Claude Reynard, 'Public order and privilege: the eighteenth-century roots of environmental regulation', *Technology and Culture*, vol. 43, no. 1, 2002, pp. 1–28.

taking statements from twelve 'convenience-inconvenience witnesses', appointed by the *parlement*, who judged the submissions made in the first inquiry. These two levels and the choice of witnesses (notables, priests and well-known artisans) bear witness to the indirect and elitist nature of consultations under the Ancien Régime.[47]

The convenience survey had the peculiarity that it placed the manufactory within its territorial context. Compared with the mercantilist arguments of the entrepreneurs or the royal council, which invoked the privileged factories' importance to the national interest, for the inhabitants, the challenge in this procedure was to secure recognition of the local interest – an intermediary between the national interest and their own individual ones. In 1776, the notables of Épinay-sur-Seine, near Paris, complained about an acid factory that had been set up without their consent. Their legitimacy to define the use of a territory that they considered to be sustained by their wealth appeared self-evident. This manufactory may 'be of some use to the state, but the town of Épinay was certainly not the place to be chosen for this establishment . . . as it is mainly full of bourgeois homes and grapevines are the main product of the territory'.[48] Given the wide range of issues addressed by the convenience survey, the procedure was transformed into an exercise in forecasting the possible future of the site. In Épinay, burghers explained that the factory would make the air unhealthy and force them to leave the town. Moreover, the workforce employed in the factory would be lost to the fields, higher wages would render the staking of the vines unviably expensive, and the winegrowers would also have to leave the village. But once the burghers and winegrowers had left, the day labourers would also find themselves destitute, as the factory would never be able to provide work for everyone. In the *commodo incommodo* reports of the eighteenth century, all arguments were admissible: environmental degradation, the lack of natural resources and manpower; but also the uselessness of the products for local activities, insufficient commercial outlets and the risk of excessive competition encouraging poor-quality products.

The subsequent development of convenience surveys is revealing. Without anticipating the next chapter too much, it should be noted that, while the 1810 decree made them compulsory for the authorisation of certain listed establishments, they now related solely to the issue of nuisances. The congruence between industry and the public good need no

---

47   AD Seine-Maritime, 1B 5527.
48   AN F12 1506.

longer be justified on a case-by-case basis. For the government, industry had become, if not an end in itself, at least an obvious vehicle for the public good. Liberal theories, and in particular Jean-Baptiste Say's law (the production of a product creates demand for another product), left fears of overproduction outmoded. From the 1830s onwards, the authorities were less concerned about diverting workers from agriculture than about leaving the dangerous classes without employment altogether. The naturally positive benefits of industry simply had to be deducted from the inconveniences, such as the fumes and bad smells it could produce locally. In the eighteenth century, the all-encompassing nature of the convenience survey enabled notables to define the use of the places they inhabited. But, in the nineteenth century, with its narrowed focus on the senses, it served above all to confine the neighbours' voice to the realm of the subjective and private harms.

5. Corseted expertise

Looking with hindsight on the police regime for regulating nuisances, medical or technical expertise is conspicuous by its absence. In Paris, it was not until the end of the eighteenth century and the arrival of police lieutenant-general Lenoir's ambitions for 'enlightened policing' that pharmacists, chemists and doctors were regularly consulted by police in matters of sanitation. The permanent expert position that Lenoir created for the pharmacist Cadet de Vaux in 1780 ('general commissioner of highways and inspector of objects of salubrity') was both a novelty and an exception.[49]

In the eighteenth century, expertise was an exclusively judicial category: the expert was appointed by a court of law, for a given case. The Parisian custom of speaking indifferently of jurors or of 'persons in the know' underlines the local and circumstantial nature of the expert's status. Expert opinion was strictly regulated, even corseted, by the adversarial judicial procedure. The parties to the case chose the experts and could dismiss them. The experts worked under their control: they were not allowed to visit premises in the absence of either party, nor conduct an investigation or experiment that was not provided for in the judgment ordering the expert appraisal.[50] The role of the expert was closer to that

---

49   Le Roux, *Le Laboratoire des pollutions industrielles*, pp. 69–108.
50   Desgodets, *Les Loix des bâtimens*, p. 32.

of an arbitrator: if the disputing parties could not agree on the facts, they could at least agree on who would define the terms of an agreement.

In cases regarding the nuisances caused by industry, the expert opinion would not appear to have had any obvious legitimacy: given that it consisted of a one-off visit, the expert could not know what living near a workshop really meant. Let us take an example. In 1782, Béville set up a vitriol factory in a *faubourg* of Rouen. His neighbours immediately went to the Normandy *parlement* to seek to have it banned. In their view, no expert appraisal was necessary: 'One should rely on the declarations and statements of the inhabitants . . . doctors and chemists cannot properly judge the dangers of the manufactory.' Indeed, 'the most judicious naturalists could not recognise them at all, because the fires of the factory are directed at will.'[51] Trade secrets, subtle variations in materials or their strength, and the various means of directing the fire, all allowed the manufacturer to deceive even the most vigilant experts. The inconsistency of the craft itself rendered the expert appraisal senseless. According to the locals, police surveillance and their concordant testimonies were the best judges. Conversely, the manufacturer demanded that the harmlessness of 'his operations be established by *technical* experiments' conducted by chemists (and this at a time when the adjective 'technical' was very rare in the industrial context). The expert appraisal ordered by the Normandy *parlement* was a compromise between the court of neighbourhood opinion and the chemists' own judgement. The experts (two doctors, two apothecaries and two gardeners) were appointed by either party. They did not write a report: their statements were taken down in minutes kept by a member of the *parlement*, in the same manner as the other witnesses.[52] Each experiment had to be approved by both parties. If it appeared to fall outside the scope of the order prescribing the expert appraisal, it had to be submitted to the court for a further ruling. Conducted in this fashion, the expert appraisal of Béville's workshop took some six months.

Scientists enrolled by the justice system as experts found it difficult to accept being mere experts for the contending parties, especially when they held prestigious positions. For example, in 1774, Lenoir asked the Faculty of Medicine to appoint three doctors to take part in an expert appraisal that was especially delicate given that it concerned the vitriol distillation workshop of Charlard, a renowned Parisian pharmacist. After making their report, the doctors launched a sharp attack on the police

51 AN F12 1507. The refutation that Béville provided to the observations of the proprietors and inhabitants of the Faubourg Saint-Sever.

52 AD Seine-Maritime, 1 B 5527.

practice of expert examinations: 'If we have delayed answering so long, you must attribute this to the obstacles, incidents, statements and counter-statements that chicanery suggests.'[53] What was most trying was having to put up with the 'oppositions, harassments and difficulties of the parties' counsel'. The appraisal was thus undermined from within by scientists who were reluctant to play the role of experts for the parties. For them, the adversarial process had become a mere squabble.

6. Customary management of environments

Police forces and *parlements* fit into a mode of veridiction that had not yet been transformed by the epistemology of the crucial experiment and fact-finding.[54] To their eyes, the legitimate decision ought not rest on a specific expert appraisal but on jurisprudence, precedents and proven guild or customary practices.

A study of the controversies surrounding the harvesting of kelp – a group of sea grasses that grow on tidal flats, whose ashes were used to make soda ash, indispensable for glassmaking – allows us to identify a mode of veridiction that is foreign to us, but which is absolutely coherent when seen in the context of eighteenth-century production.

Since the 1720s, coastal communities in the Vendée, Brittany and Normandy had accused the *soudiers*, or kelp burners, of depriving agricultural land of fertiliser (kelp was used to manure the fields), producing unhealthy fumes, causing epidemics, destroying harvests and even emptying the ocean of its fish.[55] In 1755, the *intendant* of Brittany authorised the establishment of glassworks near Lorient on condition that they imported their soda; in 1766, the Admiralty of Sables-d'Olonne forbade the burning of kelp, basing its decision on the countless complaints from the 'parish priests, syndics and inhabitants of the parishes'.[56] In Normandy, the Royal Agricultural Society of Rouen took the lead in the protest and called on the *parlement* to ban the furnaces. On 10 March 1769, the *parlement* of Normandy imposed restrictions on the harvesting and burning of kelp,

---

53   AN F12 879, 'Rapport fait à la faculté de médecine, le samedi 12 février 1774, par MM. Bellot, de La Rivière et Desessartz'.

54   Steven Shapin and Simon Schaffer, *Leviathan and the Air-Pump: Hobbes, Boyle, and the Experimental Life*, Princeton, NJ: Princeton University Press, 2017.

55   AN 127 AP 2, Le Masson du Parc papers, Letters from Mr de Gasville, Intendant of Rouen, concerning the execution of the Council ruling of 26 May 1720.

56   BM Rouen, ms Coquebert de Montbret, Y 43. Normandy's glass production doubled between 1743 and 1766. See Pierre Dardel, *Commerce, industrie et navigation à Rouen et au Havre, au XVIIIe siècle*, Rouen, 1966, p. 193.

Kelp burning[57]

citing accusations that the cutting of kelp threatened fish stocks and that smoke from kelp ovens caused epidemics and destroyed harvests.[58]

For a contemporary reader, the most remarkable thing about the *parlement*'s ruling is the absence of any reference to current scientific understanding: while it reveals a province ravaged by epidemics, lost harvests and a sea without fish, the *parlement* offered no factual evidence of the link between all these ills and the manufacturing of soda ash. The *parlement*'s evidence was of a quite different nature. After quoting at length from the ordinances and regulations dealing with kelp (the Naval Ordinance of 1681, the Royal Declarations of 1726 and 1731, which also failed to provide any proof) the text inferred the dangers of soda ash from these other texts. In the eyes of the *parlement*'s members, what they were doing was not a simple legal exercise, but a '*demonstration*'. Their recourse to precedents did not mean cutting themselves off from reality because the relationships between the different factors involved

57   Denis Diderot et Jean Le Rond d'Alembert (dir.), Encyclopédie, op. cit., Recueil de planches, Chasses, pêches, pl. XVII.

58   'Arrêt du Parlement qui ordonne que l'ordonnance de 1681, titre X, concernant la coupe du varech, ensemble la déclaration du roi du 30 mai 1731, seront exécutés selon leur forme et teneur. 10 mars 1769', *Recueil des édits registrés en la cour du parlement de Normandie*, 1774, vol. 2, p. 1123. See Gilles Denis, 'Une controverse sur la soude' in Jacques Theys and Bernard Kalaora (eds), *La Terre outragée. Les experts sont formels*, Paris: Autrement, 1992, pp. 149–57.

(fish, kelp, smoke, epidemics, plant diseases) had already been recognised in the previous century. Hence an expert appraisal would be pointless. After all, what weight could be attributed to a text written by a solitary naturalist, as compared with the repeated assent of judicial assemblies over almost a century?

And yet, the invisibility of naturalists did not imply an absence of knowledge. Through its justifications in terms of jurisprudence, the ruling, in fact, drew on practical knowledge and customary rules for managing resources. Since at least the seventeenth century, the communities on the coast of Normandy had developed institutions to handle conflicts over the use of kelp. Following a fairly consistent approach, locals would gather after Mass on the first Sunday in January to set the dates for harvesting the kelp (generally between March and April), divide up the areas between the families and appoint a 'respected and impartial' person to oversee the harvest. This communal management was aimed at preserving both the kelp and the fish: it was forbidden to uproot the plants (which had to be cut carefully), and harvesting could only take place in areas of abundance. Harvesting dates were set for after the young fish had left the kelp forests.[59]

This way of managing the supply of kelp was based on an intimate knowledge of the environment. The inhabitants of coastal communities generally practised 'small-scale fishing': during high tides, they roamed the tidal flats, armed with spades, forks and rakes to collect mussels, cockles and sand worms. They also collected fish, trapped in nets set up at strategic points along the foreshore. These specific techniques, depending on the type of catch, the time and the tides, required a wealth of knowledge about the behaviour of the prey, the suitable seabed and the dates of spawning and migration. A memorandum sent to the *parlement* by the fishermen's syndics explained that fish came to spawn in the kelp because the plants held the fish's eggs still, protected them from the tides and currents, and increased spawning density and chances of fertilisation. The young fish also found in the kelp the tiny creatures on which they

---

59   For Normandy, see Auguste Fougeroux and Mathieu Tillet, *Observations faites par ordre du roi sur les côtes normandes au sujet des effets pernicieux qui sont attribués dans le pays de Caux à la fumée du varech lorsqu'on brûle cette plante pour la réduire en soude*, 1772. and BM Rouen, ms Coquebert de Montbret, Y 43. For Brittany: Olivier Levasseur, *Les Usages de la mer dans le Trégor du XVIIIe siècle*, PhD in history, Rennes II, 2000; Emmanuelle Charpentier, *Le Littoral et les hommes. Espaces et sociétés des côtes nord de la Bretagne*, dissertation, Rennes II, 2009. For two opposing arguments related to these themes, see Garrett Hardin, 'The Tragedy of the Commons', *Science*, 162, 1968, pp. 1243–8; Ostrom, *Governing the Commons*.

fed, and protection against the backwash and predators.[60] This knowledge of the environment, largely ignored by eighteenth-century naturalists who busied themselves with classifying dead fish, was essential for the fishermen who caught them alive. In the end, it was this knowledge that formed the basis of communal management, which was taken up by customs and eventually informed royal ordinances and *parlement* rulings.

7. Botany and politics

At the end of the Ancien Régime, police, guild and customary regulation of the environment was destabilised by the process of administrative centralisation and the government's use of naturalist knowledge. Let us continue with the case of kelp. Until 1772, these sea grasses formed an indistinct group that did not yet have its own botany. Their common name was inherited from Norman custom, which defined kelp as all the things subject to 'the right of kelp', that is, to rules specifying property rights over strands. Shortly after the *parlement*'s ruling, an expedition of the best Parisian botanists, dispatched to the Normandy coast by the Académie des sciences at great expense, proposed a classification of marine plants. Eight species were meticulously described in a richly illustrated academic essay.

What was the politics behind this botany? When, in 1769, the *parlement* of Normandy issued its ruling restricting the burning of kelp, it did so in an act of opposition to the monarchy, which was encouraging glass production at the time. The master glassmakers of Normandy immediately alerted the Bureau du Commerce. In February 1770, an anonymous administrator wrote a draft document of more than a hundred pages, looking for ways to avoid a fresh crisis between the monarchy and the *parlement*. He drew up a step-by-step reasoning justifying the commissioning of an expert opinion. This process, which would seem perfectly natural a few decades later, is captured here in slow motion.

The author began by rejecting the traditional practice of consulting the interested parties, which was itself deemed to explain the successive reversals by the authorities, who were variously impressed by the particular interests of kelp burners, farmers and fishermen. The intendants' reports, which were supposed to transcend local or corporate interests, did nothing to change the situation: 'Will it be said that Messrs. the intendants

---

60   AAS, Pochette de séance, 4 July 1770.

have also been consulted? But a magistrate who is consulted consults himself.' The administrator suggested to the Bureau du Commerce that it should change its method and draw on the expertise of naturalists. By redefining the role of kelp in fish conservation, this alone could reconcile the apparently conflicting interests and reach 'a truly definitive decision'. This new power of expertise was not self-evident, and the author justifies it by referring to the physiocratic doctrine of natural law. Since, according to François Quesnay, positive laws were merely 'laws of regulation relating to the natural order', the (good) legislator should not *make* laws but content himself with knowing and promulgating the laws of nature.[61] The idea that law should be subject to nature provided the foundation for the naturalist's power to legislate.

At the government's request, the Académie des sciences thus sent two naturalists, Auguste Fougeroux and Mathieu Tillet, to the Normandy shores to study the allegations made by the coastal communities – all the better to refute them. Their knowledge of kelp at this point was zero: one even asked whether it was a plant or a concretion of tiny sea creatures. The most delicate point concerned the connection between kelp and the abundance of fish. The fishermen provided a detailed description of the relationship between the plants on the tidal flats and the life cycle of the fish, but little on the taxonomy (which fish, which kelp?). The naturalists produced an opposite description: rich on the anatomy of the organisms but with little to say about the relations between them. To demonstrate that harvesting kelp did not deplete this resource, they resorted to an experiment. A forest of kelp was divided into three contiguous sections: in one the kelp was uprooted; in the second it was cut; and in the third it was left untouched. One year later, the first section had the greatest abundance of kelp. To explain this mystery, Tillet and Fougeroux studied how kelp reproduces and referred to the new botany they were introducing. The plants used to make soda, commonly known as kelp, in fact belonged to eight species of *fucus* that have two characteristics in common: they are annuals or biennials and have no real roots but 'a kind of footing, sometimes shaped like a claw, by which they attach themselves to rocks, fallen stones, in short to bodies incapable of providing them with any food'.[62] The botanists' focus on the roots was crucial politically: if these roots did no more than hold annual plants in place, what was the point

---

61   François Quesnay, *Physiocratie, ou constitution naturelle du gouvernement*, Yverdon, 1768, vol. 2, p. 25.

62   Fougeroux and Tillet, 'Sur le varech', *Histoire de l'Académie royale des sciences*, p. 40.

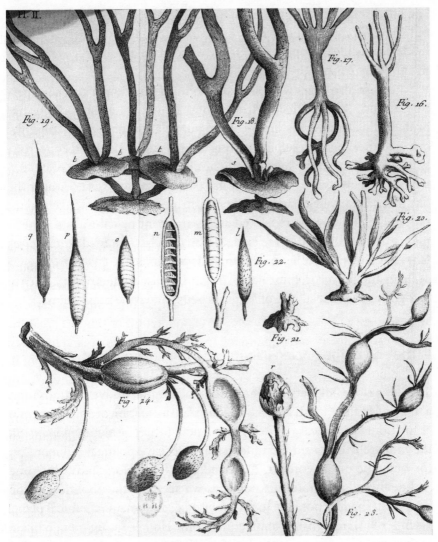

The botany of kelp in 1772[63]

of preserving them? On the contrary, it would be better to uproot them
to prevent new seeds from attaching themselves to them.

On 30 October 1772, a royal decree ratified Fougeroux and Tillet's
report: the Normandy *parlement*'s ruling from 1769 was overturned;
the kelp could be uprooted, and the coastal residents had to let the kelp
burners work on their shores. The undermining of communal management of the kelp was thus carried out indirectly – not through positive

---

63   Auguste Fougeroux et Mathieu Tillet, Second mémoire sur le varech, Histoire
de l'Académie royale des sciences, Paris, Imprimerie royale, 1772, p. 55–77.

law, but through the description of nature. Botanical knowledge cancelled out the customary uses of nature and restrictions on access to kelp. It made it possible to impose an intensive use of the resource, governed by the growing consumption of glass.

This case also bears witness to the political function of crucial experimentation. Applied to environmental issues, its epistemology has profoundly changed the *historicity* of our relationship with nature: the one-off experiment on kelp harvesting carried out by Tillet and Fougeroux supposedly proved the future resilience of this resource, even once subjected to a new usage. The correct use of the resource was no longer the one that had proven its sustainability, enshrined in centuries-old customs; rather, it could now be redefined at any moment by naturalists' experiments. The effects of the free exploitation of kelp soon made themselves felt. As early as 1811, following numerous petitions, the government gave prefects the power to limit harvesting.[64] Finally, contemporary science recognises the crucial role of seagrass beds in marine ecosystems.

8. The end of the Ancien Régime and the environment

There is a close link between the process of administrative centralisation at the end of the Ancien Régime and the transformation of environmental regulations. In the case of kelp, by silencing the Normandy *parlement*, the monarchy deprived the notables of a decisive political influence on the outcome of the conflict. A second case from Normandy, this time in Rouen will illustrate this point, as we see how the abolition of the *parlement* made it possible to set up France's first major chemical plant.

In 1768, the English manufacturer John Holker set up a huge nitric (sulphuric) acid factory in the Faubourg Saint-Sever in Rouen, intended to supply the entire national textile industry. The project was handled and encouraged by Trudaine de Montigny, the powerful director of the Bureau du commerce et des manufactures. Instead of distilling vitriols, the factory used an English technique that had never been used in France before: burning sulphur and saltpetre in a closed vessel. The factory was gigantic: 400 glass balls, each three feet in diameter, were placed in a 110-metre-long hangar at a cost of 90,000 livres. Holker received numerous privileges

---

64   In 1850, a ministerial commission pointed out the errors that the *académiciens* had made in 1772, and the 1852 law on coastal fishing authorised only one harvest per year. See 'Étude sur la législation réglementant la coupe et la récolte des herbes marines', *Revue maritime et coloniale*, 63, 1879, pp. 6–14.

from the Bureau du commerce, including exemption from taxes and a bonus of ten livres per quintal of acid produced. Each year, he received between 3,000 and 5,000 livres in subsidies.[65]

The factory also had particularly ill effects. A police *commissaire* noted: 'The factory is not only harmful to plants, but also to human temperament and health.'[66] Its opponents were both numerous and influential: noblemen with a passion for agronomy and botany; members of the *parlement*; jurists; market gardeners who used land valuable because of its proximity to Rouen; and dyers whose fabrics were damaged by the acid vapours. In 1770, thanks to the support of the Normandy *parlement*, the opponents were on the verge of victory.

But the abolition of the *parlement* by Chancellor Maupeou in 1771 and its replacement by a Higher Council controlled by the monarchy completely changed the balance of power. Trudaine de Montigny, at Holker's request, had the government rule on the procedure in the Higher Council; and leaving nothing to chance, the matter was referred to the Royal Council. A ruling in 1774 prohibited the factory's opponents from 'disrupting its activities' on pain of a 1,000-livre fine. Holker's victory was complete. The letters of nobility he received a month after the ruling emphasised his industrial success: 'Chemistry has revealed to him the secret of the composition of vitriol oil, with which he is now able to supply our kingdom.'[67]

The Holker affair set a fundamental precedent: the survival of France's first major chemical manufacturer depended on the abolition of a *parlement*. In the eyes of the government, factories should no longer be dependent on the vagaries of judicial rulings. Maupeou explained that the aim of abolishing the *parlement* of Normandy was 'to put an end to the spirit of chicanery that often drags out trials [and to] give trade all the activity of which it is capable'.[68] Holker's trial was part of this pro-industrialist and anti-judicial policy. Trudaine de Montigny justified his intervention in the case by pointing to the scale of the capital involved: 'As the manufactory in question was established at great expense, it deserves

65   AN F12 1506; John Graham Smith, *The Origins and Early Development of the Heavy Chemical Industry in France*, Oxford: Clarendon Press, 1979, pp. 7–11. Between 1740 and 1789, the monarchy paid out 5.5 million livres in subsidies. See Philippe Minard, *La Fortune du colbertisme. État et industrie dans la France des Lumières*, Paris: Fayard, 1998, ch. 7.

66   AN F12 879.

67   AD Seine-Maritime, C2312 fol. 174–5.

68   Cited in Jules Flammermont, *Le Chancelier Maupeou et les parlements*, Paris: Picard, 1883, p. 460.

protection . . . we cannot dispense with showing consideration for those who are its entrepreneurs.'[69]

The Holker case also highlighted the gap between inconvenience and insalubrity, between the senses and health – a gap in which the expertise and power of hygienists would flourish in the early nineteenth century. In 1773, while the Holker case was pending before the Royal Council, the chemist Louis Bernard Guyton de Morveau revolutionised 'air purification' processes: the vapours given off by a mixture of salt and sulphuric acid instantly freed Dijon cathedral of the putrid smell of its sepulchral cellars.[70] The technique was approved by the Academy, and Holker sent an article on the experiment to the Royal Council. This was an extraordinary boon for the chemical industry: now, far from something unhealthy, it remedied the infections of the air!

The change in doctrine on acid fumes came rapidly. As late as 1768, the Faculty of Medicine, issuing an opinion on a Parisian nitric acid distillation workshop, judged them to be 'dangerous and very inimical to the chest'. Distance did not lessen the danger, as it had to do with *particles* which travelled without changing in character. Logically enough, the Paris *parlement* ordered that the workshop be closed. In 1774, the same faculty, consulted by Lenoir about the nitric acid workshop run by the pharmacist Charlard, overturned its 1768 decree. The fumes from the workshops were no longer intangible particles, but *effluents* that dissolved, mixed and cancelled each other out: 'These vapours . . . could be *inconvenient*, but in no way *harmful*.'[71] The programme for the correction of the atmosphere using chemicals made it possible to dismiss olfaction as a yardstick of healthy conditions.

In moving from the distillation of vitriols to the combustion of sulphur in a closed vessel, acid production intersected with the history of the so-called chemical revolution. Let us read the famous sealed envelope that Antoine Lavoisier submitted to the Académie des sciences on 1 November 1772, introducing his theory of combustion:

> About eight days ago, I discovered that when sulphur is burned, far from losing weight, it instead gains weight; this means that from a pound of

---

69 AN F12 879, Trudaine à M. de Perchel, procureur général du Conseil supérieur à Rouen, 23 March 1773.

70 Louis Bernard Guyton de Morveau, *Traité des moyens de désinfecter l'air, de prévenir la contagion et d'en arrêter le progrès*, Paris: Bernard, 1801.

71 AN F12 879, 'Rapport fait à la faculté de médecine par MM. Bellot, de La Rivière et Desessartz pour examiner le laboratoire du sieur Charlard', 1774, p. 16. This concerned Jacquet's workshop.

sulphur one can draw much more than a pound of vitriolic acid . . . This increase in weight comes from the prodigious quantity of air that is fixed during combustion.[72]

In a sense, Lavoisier was simply rediscovering the vitriolic acid manufacturers' own practice. At the end of the 1760s, they were the experts in sulphur combustion. They had a quantitative knowledge of it, since the ratio between sulphur burnt and acid produced determined their profits. Demachy had conducted experiments to determine the weight of sulphur and saltpetre according to the desired quantity of acid. Indeed, Lavoisier was closely involved with the main players in the nascent vitriolic acid industry. He was familiar with Demachy's work, on which he reported to the Académie.[73] Together with John Holker's son, he also attended the chemistry lectures given by Guillaume-François Rouelle at the Jardin Royal des Plantes. Last, but not least, Lavoisier and Holker shared the same protector: Trudaine de Montigny, the director of the Bureau du Commerce, who owned a large chemistry laboratory where Lavoisier conducted his experiments. In 1774, Lavoisier dedicated his first book to Trudaine de Montigny: 'You have more than once guided me in the choice of experiments, [you] have often enlightened me on their consequences . . . what motives for offering this essay to you!' It is possible that Lavoisier's work in 1772 fitted not only into his theoretical research into the combustion and decomposition of air, but also into the manufacturing programme for vitriolic acid production that his protector Trudaine de Montigny had been ardently encouraging since 1768.[74]

Lavoisier's new chemistry played a key ideological role in the transformation of environmental regulation in France: a factory could henceforth be conceived as a vast laboratory experiment. Knowing what masses of chemicals would be involved in a reaction made it possible to envisage a perfect process in which all the reagents would be transformed without residue and thus without pollution. Everything would be transformed, without anything being created or anything lost, especially to the atmosphere. An inevitable phenomenon in small-scale chemical production, leakage was now seen as a loss of product and a financial waste.

---

72   BNF ms fr. 12306, Macquer correspondence, fol. 123–4.

73   Antoine Lavoisier, *Opuscules physiques et chymiques*, Paris: Durand, 1774, dedication.

74   Compare Bernadette Bensaude-Vincent, *Lavoisier*, Paris: Flammarion, 1993, pp. 56–7; Frederic Lawrence Holmes, *Antoine Lavoisier: The Next Crucial Year*, Princeton, NJ: Princeton University Press, 1998, p. 39.

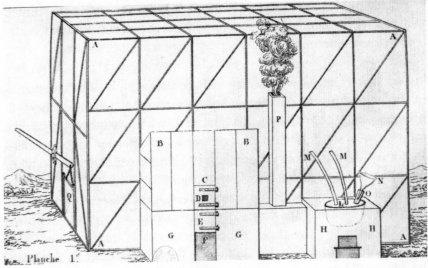

Lead chamber[75]

The lead chamber was a crucial innovation in this early industrial utopia of control and impermeability. In Rouen from 1772, and later in Paris, Lille and Montpellier, acid factories equipped themselves with vast metal vessels to burn large quantities of sulphur and saltpetre. Compared with the small-scale production of vitriol oil by artisans described by Demachy, this device instead centred the production process on a single technical object and downgraded both the workers' efforts and the infinite variability of processes and materials. As well as confining the material, the lead chamber was also a technique of social confinement: production was cut off from know-how and rules of the trade. A new class of experts could now purport to perfect the production process. The lead chamber made it possible to imagine a technical solution to the environmental problem. It was no longer necessary to control myriad artisanal gestures or to ensure the impermeability of hundreds of stoneware vessels sealed with dung. Rather, it would suffice to control the operation of a single metallic and hermetic object. This new method held out the hope of resolving environmental conflicts by honing the production process.

In 1774, a dyer in Rouen who had installed a lead chamber received complaints from his neighbours about the acid vapours. La Follie, a chemist and member of the Académie de Rouen, came to his aid and proposed an improvement that would be adopted by all manufacturers: to condense the sulphurous vapours and prevent them from escaping,

---

75   B. Rougier, Mémoire sur la fabrication de la soude artificielle, Mémoires de l'Académie de Marseille, vol. 10, 1812, p. 117.

he sprayed water into the lead chamber. La Follie made it a fundamental point that profit and the environment were not in contradiction and could be reconciled by technique: 'The less loss there is for the manufacturer, the less the neighbours will be exposed to inconveniences.'[76] In Lille in 1781, doctors sent by the aldermen to report on a lead chamber also proposed a technical improvement. Since acid vapours escaped from the chamber through a single opening, all that was needed to absorb them was to instal tanks full of water at this exit point. The lead chamber and other similar industrial devices transformed the types of regulation that could be contemplated. They are essential to understanding how technology, that is the study and improvement of techniques, suddenly assumed a natural role in the government of production at the beginning of the nineteenth century.

The notions of *technology* and *technical improvement* destabilised the Ancien Régime of environmental regulation. The normative texts of police and *parlements* had regarded trades and productive activities as stable entities whose legal status would not require regular revision. For example, the countless ordinances on tanners that had followed one another since the Middle Ages, relegating them to the lower reaches of rivers, considered tanning to be a non-evolving activity. Its processes, and the substances it used and rejected (the eternal 'treads and notches' of police ordinances) were considered inexorable.[77] Similarly, in the 1769 ruling of the Normandy *parlement*, 'the soda factory' was assumed to be unchangeable, thus justifying the eternal validity of old texts. For the political and judicial institutions of the Ancien Régime, precedent was the bearer of experience and exemplarity, on which the enactment of contemporary norms had to be based. Sound policing was based on a tried and tested precedent. This respect for the past led the magistrates to give extraordinary attention to even the oldest of texts. Most of Delamare's *Traité de la police* (1705) was a compilation of ordinances dating back to the Middle Ages. To justify the need to keep 'livestock that cause infection' out of the towns, Delamare unearthed the 'Great Ordinances' of Louis IX that were already half a millennium old rather than quote contemporary medical literature on the subject.[78] The police and *parlements* of the Ancien Régime considered nature and productive worlds

---

76   La Follie, 'Réflexions sur une nouvelle méthode pour extraire en grand l'acide du soufre sans incommoder ses voisins', *Observations sur la physique, sur l'histoire naturelle et sur les arts*, Paris: Chez Ruault, 1774, vol. 4, p. 336.

77   Delamare, *Traité de la police*, vol. 1, pp. 553-7.

78   Ibid., p. 539.

through a different historicity from that to which we are today accustomed by scientific expertise and administrative practice. Guild jurisprudence, customary practices, judicial precedents and the exemplary nature of the past gave historical consistency to decisions concerning the worlds of production, the uses of nature and their sustainability.[79]

At the end of the eighteenth century and the beginning of the nineteenth, a new regime established itself. Now, technical innovations and naturalist knowledges would take precedence over the knowledges embodied in the customs and communities of the trades. In 1810, the drafters of the foundational decree on classified establishments claimed that they had to formulate a new code of nuisances precisely to take account of the many technical innovations of the late eighteenth century. The decree of 1810 marked a break with the Ancien Régime's practice of heaping together textual authorities (or, as eighteenth-century jurists saw it, drawing on consolidated knowledge). Rather, it made a *tabula rasa* introducing a normative practice based on contemporary scholarly expertise rather than on past judicial experience. The rising role of expert opinion in defining sound production practices, the proper uses of nature, and what is dangerous and unhealthy, was part of a wider change in the historicity of government practices. From now on, the truth would come from the present.[80]

79   Alice Ingold, 'To historicize or naturalize nature: hydraulic communities and administrative states in nineteenth-century Europe', *French Historical Studies*, vol. 32, no. 3, 2009, pp. 385–417.

80   Reinhart Koselleck, *Futures Past: On the Semantics of Historical Time*, New York: Columbia University Press, 2004.

4

The Birth of the Polluter Pays Principle

From the 1800s, a new line appeared in the accounts books of polluting firms: damages, compensation or legal costs. These terms encapsulated what had happened to the environment at the turn of the eighteenth and nineteenth centuries. From its status under the Ancien Régime as a common good that determined human health and needed policing, the environment became the subject of financial transactions, compensating for the damage bemoaned by the industrialists' neighbours.

To understand this, we can take the example of a large chemical plant established in Salindres, near Nîmes, in 1855. Right from its foundation, the company paid annual indemnities to farmers within a three-kilometre radius. It was only in 1871, when the industrialists refused to pay the compensation, that the government was obliged to send in its experts. Although the authorities intervened very rarely and in a timid fashion, it was indeed the existence of compensation, paid regularly by mutual agreement, or else arbitrated by the civil courts, that prevented the environmental conflicts of the Industrial Revolution from intensifying.

The government and the civil courts represented two sides of one liberal system of environmental regulation: by authorising classified establishments in accordance with the procedure set out by the 1810 decree (*commodo incommodo* inquiries and expert appraisals), the authorities guaranteed their continued existence, notwithstanding any objections from neighbours. Since the neighbours could have no hope of the factory being closed down, they were left with no choice but to turn to the civil courts to seek compensation. In making industrialists pay a price for pollution, the civil courts were supposed to provide the financial incentives for reducing emissions.

How did the *circumfusa* become a matter of accounting? By what historical process could the environment – the central focus of the Ancien Régime police force, and the essential object of the governance of health and the population – be subjected to simple financial compensations and private arrangements?

1. The political economy of the 1810 decree

Historians have already studied the regulation of classified establishments in the period following the French Revolution.[1] Some centre their analysis on the history of the olfaction which structured risk perception, and others on the chemical industry as 'judge and jury' – emphasising that the biggest polluters, Chaptal first among them, inspired the decree. Yet, in each case, they overlook one essential aspect: that the new regulation was, above all, a liberalisation and commodification of the environment, adapted to the emergence of industrial capitalism. It was part of the great transformation described by Karl Polanyi: 'Machine production in a commercial society involves, in effect, no less a transformation than that of the natural and human substance of society into commodities.'[2]

Post-revolutionary environmental regulation began with a critique of the police. In their seminal 1804 report on insalubrious factories, Chaptal and Guyton de Morveau were perfectly explicit: 'As long as the fate of factories is not assured . . . as long as the manufacturer's fortune or ruin rests in the hand of a *simple police magistrate*, how can he conceivably be so imprudent as to engage in undertakings of this nature?'[3] The demands of capital no longer tolerated the uncertainties of policing.

Chaptal was a great advocate of industrial capitalism. As early as 1798, in a book whose ambitions belied its modest title, he proposed a new political order centred on the entrepreneur. In Chaptal's view, all society had to be reformed so that industrialists could make proper risk calculations. The fickle policies of the Ancien Régime had discouraged investment; instead, capitalists had to be guaranteed 'against events', their property and supplies had to be secured, and a stable customs and tariffs

---

1   For an overview of the 1810 decree, see section 4 of this chapter: 'Deceptive regulation'.

2   Karl Polanyi, *The Great Transformation: The Political and Economic Origins of Our Time*, Boston: Beacon Press, 2001 (1944).

3   Jean-Antoine Chaptal and Louis-Bernard Guyton de Morveau, 'Rapport sur les fabriques dont le voisinage peut être nuisible à la santé', *Procès-verbaux des séances de l'académie*, Hendaye: Imprimerie de l'observatoire d'Abbadia, vol. 4, 17 December 1804.

policy had to be guaranteed.[4] The society of which Chaptal dreamt was that of the industrial *fait accompli*, where industry stood outside politics. Protecting it was imperative: 'Whatever manufacturing industry is established, the government owes it protection: *once it exists, there is no longer any question of examining whether it was advantageous to introduce it.*' If a government had to choose between protecting manufacturers or agriculture, it must choose the former, because 'agricultors may have setbacks in their operations, but their capital remains, while for the manufacturer all is lost'.[5]

The 1804 report fitted right into this programme. Ancien Régime–era environmental regulation was blamed for producing a state of uncertainty, an *eternal indeterminacy* that disincentivised entrepreneurs. They ought to be freed from the police's grasp; the administration had to 'set limits that no longer leave anything to the arbitrary choice of the magistrate, and which mark out the remit within which the manufacturer may freely and assuredly practice his industry'. The 1810 decree gave this project material force: the administration (the minister of the interior, the Council of State and the prefectures) subjected factories to rigorous authorisation procedures and, in exchange for this, guaranteed their continued existence. In 1819, Chaptal looked back with satisfaction on the legislation which he had helped to implement:

> Since the chemical trades made astonishing progress . . . complaints arose from all sides and the courts were making rulings in an arbitrary manner . . . The fate of factories was, so to speak, at the mercy of an anxious neighbour . . . Fixed rules had to be established in this respect: the decree of 15 October 1810 contains rather sound legislation on this matter.[6]

The new environmental regulations were thus an adaptation to the needs of the French chemicals industry. Between 1790 and 1810, as a result of the revolutionary wars and the continental blockade, this industry changed radically in scale. While there were only a dozen sulphuric acid factories in 1789, more than forty were set up during the Revolution and the Empire, in all the major urban centres (including Paris, Lyon, Rouen, Marseille, Montpellier, Nantes, Mulhouse, Nancy, Strasbourg, Amiens),

---

4   Jean-Antoine Chaptal, *Essai sur le perfectionnement des arts chimiques*, Paris: Crapelet, 1798, p. 51, and *De l'industrie française*, Paris: Renouard, 1819, vol. 2, p. 443.
5   Ibid., p. 418.
6   Ibid., p. 369.

without concern for where in the city they should be located.[7] Something else changed too, for the new artificial soda factories (around fifty were set up at the end of the 1800s, mainly in Paris, Rouen and Marseille) produced unprecedented levels of pollution. The production of two tonnes of soda using the Leblanc process (based on salt and sulphuric acid) gave off one tonne of hydrochloric acid vapour, which corroded everything in the vicinity. In 1818, the hinterland of Marseille produced 15,000 tonnes of soda annually, and 30,000 ten years later.[8] The revolutionary suspension of traditional forms of regulation (the abolition of guilds and *parlements*) and the massive development of industrial chemistry were the two main driving forces behind the new regime of environmental regulation.

2. The state construction of the chemical market

During the Revolution and the Empire, chemists eagerly followed the shortages and price hikes that swept across Europe, in line with the spread of battles and blockades. In 1794, Chaptal was ruined by the closure of his factory in Montpellier and the devaluation of the *assignats*. He rebuilt his fortune by selling acid to factories in Catalonia, which were deprived of British products because of the war: 'As I had no competitors, I set a very high price for my products. In one year I made a profit of 350,000 francs.'[9] The war against Spain in 1808, which brought an embargo on natural soda ash imports, enabled chemists to rake in vast profits. In Paris, the price of a quintal of soda plant rose from 60 to 280 francs in a few weeks. In 1809, in Marseille, soap manufacturers were paying more than 300 francs a quintal.[10] Fortunes were made in just a few months: 'Messrs Gautier and Barrera, who have been making [soda] for over a year, have earned several millions. Natural soda sells for 160 and artificial soda for 80. God forbid that the prices change.'[11]

---

7   John Graham Smith, *The Origins and Early Development of the Heavy Chemical Industry in France*, Oxford: Clarendon Press, 1979; Xavier Daumalin, *Du Sel au pétrole*, Marseille: Tacussel, 2003.

8   The Leblanc process produced a vast amount of waste: in the 1820s, to produce one unit of soda ash required 1.5 units of coal, 1 of chalk, 0.7 of sulphuric acid and 0.7 of sea salt. See Christophe de Ville-Neuve, *Statistique des Bouches-du-Rhône*, Marseille: Ricard, 1824, pp. 787–95. In 1862, a House of Lords report estimated that for every 280,000 tonnes of soda ash in Britain, 3,873,000 tonnes of waste (hydrochloric acid and soda dregs) had to be discharged.

9   Jean-Antoine Chaptal, *Mes souvenirs sur Napoléon*, Paris: Plon, 1893, p. 52.

10   Smith, *The Origins and Early Development of the Heavy Chemical Industry*, pp. 245, 273.

11   Letter from Philippe Girard in 1809, quoted by Daumalin, *Du Sel au pétrole*, p. 31.

But, in peacetime, there was no market for artificial soda. Its commercial outlets depended on customs laws and tax policy: with the reintroduction in 1806 of a tax on salt (20 francs per quintal), the price of the salt used in the production of artificial soda was *higher* than the price of natural Spanish soda. From 1806 onwards, chemists were thus constantly lobbying to obtain tax exemptions. Chaptal, in the Conseil d'État, and Claude-Anthelme Costaz, at the Bureau des manufactures, spoke up for the soda manufacturers' cause, and on 13 August 1809, Napoleon agreed to their request.[12] Immediately, soda works in Paris, Rouen and Marseille ramped up production to take advantage of the favourable prices.[13] Alas, in 1810, the resumption of trade with Spain and the return of natural soda put the new industry under threat. The soda manufacturers invoked the protection granted in 1809 and the investment it had generated to call for a ban on imports of natural soda. The ever-sympathetic government acceded to their request in July 1810.[14]

In addition to extraordinarily favourable customs policy, the authorities and their experts helped to create a demand for artificial soda. At the end of the eighteenth century, there were as many types of soda ash as there were sources and uses: Normandy kelp was used by glassmakers because it contained materials from the soil useful for vitrification and Alicante soda ash for dyers.[15] Artificial soda made with the Leblanc process disrupted this economy: soap manufacturers complained of the sulphides that gave their products a nauseating odour, dyers noted damage to fabric, and glassmakers were unable to obtain white glass. For artisans, artificial soda was simply an ersatz that would disappear once European trade resumed. In 1810, they petitioned the chambers of commerce to end the ban on the importation of natural sodas.

The alkalimetric dosing technique invented by the Rouen chemist Descroizilles in 1806 played a fundamental role in teaching artisans how to use the new chemical products. Alkalimetry measured (and invented)

---

12   AN F12 2245, Extrait du registre des délibérations du Conseil d'État, 13 June 1806.

13   To prevent salt delivered free of duties from being resold, there was an employee of the tax authorities residing in each factory. He was required to observe their operations and to check that all the salt had indeed been instilled with acid. The problems with pollution control did not, therefore, owe to any lack of administrative oversight. See C. F. de Saint-Genis, *Essai sur l'établissement et la surveillance des fabriques de soude factice*, Marseille: Allègre, 1829.

14   AD Seine-Maritime, 5M 316, Lettre des fabricants de soude de la ville de Rouen à M. le préfet.

15   5 M 766, Pelletan, *Rapport sur la soude envoyé à l'Académie de Rouen*, 12 June 1805.

a concept that had not hitherto existed in the artisans' world, that of the *strength* of soda. It reduced all the natural sodas with their varied qualities to a single metric. It set a new standard for the market, which was no longer based on origins and qualities, but on concentrations and strengths. Finally, it provided an essential commercial argument, as it appeared that artificial soda always had a higher alkali content than natural soda.

To convince artisans to give up using natural soda, chemists commissioned by the political authorities explained that they simply needed to add various substances in order to recreate the properties of the natural soda. In Rouen, the prefect, who was a close associate of Chaptal, oversaw the development of the chemical industry. In 1803, he set up a course in chemistry applied to the trades, which taught dyers in Rouen how to use artificial soda. The government also published an official instruction manual to convince glassmakers to use chemical soda, which had to be softened by adding silica. For soap manufacturers, the process was reversed: in this case, the sulphides in the artificial soda had to be removed.[16]

According to the soda makers, these minor hassles were easily offset by the homogeneity of the chemical product. Selecting a natural soda with hidden qualities was a delicate question. For instance, a mistaken choice could spoil a dye bath. Conversely, artificial soda was deemed 'consistently successful', allowing for a 'regularity of processes and doses'.[17] Dosages and pure products abruptly devalued the knowledge on the quality of materials (their taste, smell or consistency) that had granted the artisan chemist such an eminent role. In England, the great dosing specialist was none other than Andrew Ure, well known for having formulated, in his *Philosophy of Manufactures* (1835), the first major theory of the factory system and the disembodiment of artisan knowledge by the machines.[18]

---

16   C. Pajot-Decharmes, *Instruction sur l'emploi des soudes factices indigènes à l'usage des verriers*, 1810. In Marseille, the chemist Laurens invented a desulphurisation process to encourage soap-makers to abandon the use of Spanish soda. See archives de l'Académie des sciences de Marseille, 'Recherches sur les savons du commerce, suivies de quelques observations relatives aux moyens de détruire les sulfures contenues dans lessives des soudes artificielles', 1811.

17   Pelletan, 'Rapport sur la soude.'

18   Andrew Ure, *Dictionnaire de chimie*, Paris: Leblanc, 1822, vol. 1, pp. xx–xxv. Ure invented not only an 'alkalimeter' but also an 'acidometer', an 'indigometer' and a 'bleachometer'.

3. Capital pollution

How did Napoleon's administration manage the environmental consequences of its industrial policy? We need to distinguish between two
cases: sulphuric acid factories, on the one hand, and soda works on the
other. Each called for different strategies. The former tied up considerable
capital: a lead chamber required 5 tonnes of metal and cost around 30,000
francs.[19] In 1840, a bankrupt chemical products factory was valued at
400,000 francs, including 200,000 for the lead chambers.[20] The Chaptal
fils and Berthollet fils chemical plant at Plan d'Aren near Marseille had
capital of 1,200,000 francs. For the government, it was unthinkable to
order the relocation of such facilities.

Two complementary solutions were pursued. The first was to educate
neighbours to change their perception of the risks, to make them accept
the new olfactory experience represented by acidity – and take it for a sign
of healthiness. The context was favourable to this: we have seen how the
purification of the atmosphere by acid fumigations promoted by Guyton
de Morveau allowed chemical plants to be presented as healthy.[21] In their
1804 report, Chaptal and Guyton distinguished between two classes of
factories: on the one hand, workshops that putrefied vegetable or animal
substances (such as hemp retting, poudrette, gut factories, butcheries
and tallow foundries); and on the other, the new chemical factories. The
former were said to give off 'insipid, foul-smelling miasmas that are very
dangerous to breathe', and should thus be kept away from homes. The
latter, on the other hand, were perfectly harmless and could even purify
the atmosphere. This report was immediately cited by the minister of
the interior to order the maintenance of the sulphuric acid plants in the
Saint-Sever district, despite the unanimous complaints of the residents:
'It is for want of better enlightenment [that they] claim that the emissions
from acid factories are harmful to human health and plant vegetation.
This is the opinion of the Institut.'[22]

---

19   *Mémoire d'Estienne et Jallabert, fabricants de produits chimiques à la Guillotière*,
1840. At the time, a worker's daily salary ranged between 1 and 3 francs.

20   AD Bouches-du-Rhône 410 U 50, Rapport d'experts sur la faillite de Grimes
et Cie, mai 1840.

21   Guyton, *Traité des moyens de désinfecter l'air* was reissued three times between
1801 and 1805. In 1803, Chaptal sent a circular to the prefects recommending this
technique.

22   AD Seine-Maritime, 5 M 763, lettre du ministre de l'Intérieur au préfet de la
Seine-Inférieure, 11 October 1805.

The second solution was technological improvement, or the idea of it. At the beginning of the century, the rapid increase in the yield from lead chambers (more and more acid was being produced with the same quantities of sulphur and nitrate) raised hopes for the advent of a *perfect* industry that would not lose any nitrous or sulphurous gases into the atmosphere. According to Chaptal, the acid industry 'has reached perfection, since according to the known analysis of the acid obtained, it is proven that not a single atom of sulphur is lost in the process'.[23] This conception of the factory as an input/output system, in which the manufacturer maximised yields, was the logical basis for a liberal regulatory system: since the loss of material now meant financial loss, the entrepreneur would surely reduce pollution all by himself.

The Conseil de salubrité de Paris (a misnomer) was founded in 1802 in order to encourage industry, to help the prefect of police authorise workshops and rebut neighbours' opposition. It openly promoted techno-liberal ideology. For instance, its founder, Charles-Louis Cadet de Gassicourt, concluded a report on the factory that Chaptal had established in Neuilly in the following terms: 'Given that it is in *the entrepreneur's interest* not to lose any vapours, the factory . . . cannot give rise to any reasonable complaint. If neighbouring proprietors were subsequently inconvenienced by it, this would be entirely the fault of the workers.'[24] Leakage, a normal phenomenon in the chemicals industry of the eighteenth century, was now seen as an incident caused by the mistakes of the workers, who thus needed to be disciplined.

This techno-liberal discourse met with many rebuttals. First, the transformation of sulphur into sulphuric acid was still a poorly understood process: manufacturers in the 1810s–1830s thus spoke of 'soft' or 'hard' lead chambers according to their effectiveness, or even of a mysterious 'chamber disease' that prevented sulphuric acid from forming.[25] Second, the condensation of the vapours posed considerable technical problems, and most importantly, slowed production. The high fixed costs (tied-up capital and workers' wages) led entrepreneurs to mass produce at breakneck speed, even if it meant losing out on output.

---

23   Chaptal, *De l'Industrie française*, 2, p. 65.

24   APP, RCS, 16 September 1811. See also Arvers, *Mémoire sur les fabriques d'acides dans le département de la Seine-Inférieure, lu à la Société d'émulation de Rouen*, 9 June 1817.

25   AD Seine-Maritime, 5 M 316, Rapport de Vitalis au préfet, 12 January 1810; Alexandre Baudrimont et al., *Dictionnaire de l'industrie manufacturière, commerciale et agricole*, Paris: Baillière, 1833, vol. 1, p. 125.

Finally, the government's desire to 'perfect' the industrial process came up against the problem of manufacturing secrets: manufacturers with the best yields, and therefore the least polluting plants, had no interest in divulging their methods.[26] A well-performing lead chamber offered a vast competitive advantage. In 1820, Chaptal, whose plant at Les Ternes near Paris had the best yields, entered into a treaty with the Plan d'Aren company: it would pass on the secret of its lead chamber in exchange for one-third of the profits. The savings on sulphur and saltpetre amounted to some 30,000 francs a year.[27] The Plan d'Aren company then did everything in its power to protect this advantage: the lead chambers were locked up by the workers in a shed called 'the secret'. It was locked and entry was forbidden except to trusted employees.[28]

For the opponents of the acid factories, pollution was intrinsic to chemical capitalism. Chaptal's neighbours at les Ternes complained: 'Insalubrity is not an accidental inconvenience . . . The showy rhetoric about the progress of science, the perfection of processes, and the infallibility of preservatives, no longer convinces anyone.'[29]

When it came to the artificial soda factories using the Leblanc process, the industrialists' parrying efforts were worthless. Wherever such establishments were founded, complaints immediately poured in. The heavy clouds of hydrochloric acid rising from the furnaces produced such massive and damaging pollution that the exculpation strategies described above appeared derisory. How could the neighbours of the soda works, who saw their crops devastated, their trees withering or their metals rusting, ever be convinced that the suffocating fumes of hydrochloric acid actually made the air healthier?

In autumn 1809, the soda makers rapidly produced a lot of soda in order to take advantage of the high prices. They were well aware of the damage they were causing, but they ignored it: they had to make their expensive lead chambers pay off. While acid was not worth much, soda was worth a fortune. A manufacturer waiting for authorisation from the prefect before starting production requested a 'prompt response: you

---

26   AD Seine-Maritime, 5 M 763, Rapport sur l'établissement des fabriques d'acide sulfurique dans le voisinage des habitations, 1806.

27   AN F12 6728, Traité du 5 juillet 1820.

28   ASM, Mémoire pour la compagnie anonyme du plan d'Aren contre le sieur Castellan, 1829.

29   *Mémoire au Conseil d'État pour les habitants des Ternes contre M. le sénateur Chaptal*, 1811, p. 39–57.

know better than anyone the cost of time in matters of manufacturing.[30] These first hastily built workshops would move around according to complaints and prefects' decisions. In Bergheim, near Colmar, the mayor asked the gendarmes to close down a soda workshop consisting of two 'shacks made of old boards'.[31] In Paris, Charles Marc and Costel – forced to leave Gentilly after only a few weeks' production – offered to set up their workshops near the highway at Montfaucon. Nicolas Deyeux, a member of the Paris Conseil de salubrité, approved the project: 'Marc and Costel would do a great job of a sort of Guytonian fumigation.' After a few months' operations, the complaints were so vehement that the Conseil de salubrité was forced to back down. During a visit, its members were caught in a cloud of hydrochloric acid: 'All the persons who were submerged in this acidic atmosphere were seized with a stubborn cough and could scarcely breathe.'[32] The prefect closed the factory down. Gauthier, Barrera and Darcet, showing more caution, set up their workshop in La Folie, a deserted district to the west of Nanterre. There were no dwellings within two kilometres. But this was to no avail: the wind carried the acidic vapours a great distance, and the mayor of Nanterre called for operations to be suspended.[33]

The situation was graver still in Marseille: four small workshops prospered in the heart of the city, and five more were planned.[34] The city's medical society decreed that living next to soda factories was 'harmful to health' and advised that they should be moved a mile away from inhabited and cultivated areas.[35] The report was published in the *Journal de Marseille*, and the mayor ordered the closure of two of the workshops.[36] At the same time, in Béziers, a soda manufacturer was forced to cease production.[37] In Rouen, the prefect received petitions with up to 300 signatures. The police commissioner painted a catastrophic picture of the Saint-Sever neighbourhood, where the factories were located: the vegetation was burnt out, the inhabitants had to leave their homes and the textile workers fled their factories, which were drowned in the acidic fog.[38]

---

30   AD Seine-Maritime, 6 M 766, Lefrançois au préfet, 9 October 1809.
31   AD Haut-Rhin, 5 M 97.
32   APP, RCS, 3 September 1809 and 4 December 1810.
33   APP, RCS, 26 December 1808, 30 September 1809 and 4 May 1810.
34   AN F12 2245, 'Ateliers de soude dans Marseille', June 1810.
35   Public meeting of the Marseille medical society, held in the museum's main hall, 25 November 1810.
36   François-Emmanuel Fodéré, *Traité de médecine légale et d'hygiène publique ou de police de santé*, Paris: Mame, 1813, vol. 6, p. 323.
37   AD Hérault, 5 M 1014, Lettre du préfet Nogaret, 29 December 1809.
38   AD Seine-Maritime, 5 M 763, Rapport relative aux fabriques de produits chimiques, 11 January 1810.

The soda makers had to leave the suburbs. They moved into the hin-terland of Marseille, and particularly to the village of Septèmes. In Rouen, Prefect Savoye-Rollin moved them to Bruyères-Saint-Julien (the lead chambers for sulphuric acid remained in Saint-Sever).[39] The prefect pro-posed a financial solution to the keeper of waters and forests, who was concerned about the effect on the woods: the soda makers would com-pensate the administration for each hectare of woodland 'recognised to have been struck by sterility'. Peasants were subjected to the same system. To avoid any disputes, the entrepreneurs would pay these indemnities collectively, in proportion to their production.

The introduction of regulations based on compensation for damage as early as 1809 was linked to the sheer amount of capital mobilised for chemicals production. Condensing vapours by bubbling would have been possible, but would have required working in a closed vessel, slowly and in small volumes. To make the capital invested pay off, it was necessary to produce quickly and in large quantities. According to Savoye-Rollin, compensation was the only solution if the soda makers were to 'be able to operate freely and economically'. Similarly, despite the peasants' insis-tent demands, the soda makers systematically refused to stop production during fruiting or harvest periods, as their factories had to run at full capacity to be profitable. The high fixed costs associated with the lead chambers made any productive compromise difficult. No matter how much sulphuric acid and soda were produced, lead chambers deterio-rated: 'Rest is synonymous with destruction: everything rusts, everything oxidises.'[40] Whereas in the eighteenth century, regulations prohibited the burning of kelp during fruiting or harvesting periods, or when the wind was blowing from the sea, the logic of capital was now antithetical to this fine management of temporalities.

4. Deceptive regulation

The decisive decree of 15 October 1810 was drawn up in a context of unrest stirred up by the soda industry. The interior minister Montalivet first com-missioned a report from the Institute. Its urgency was manifest: the report was requested on 2 October 1809, and by the 23rd the minister was already getting impatient. The report was delivered on the 30th. Although it was

---

39   Ibid., 5 M 316.
40   AN F12 2243, *Mémoire des fabricants de soude de Marseille en réfutation d'une pétition de quelques habitants du département de l'Aude*, 1824, p. 27.

general in scope, the Leblanc process was on everyone's minds. Deyeux, who was the main drafter of the report (along with Chaptal, Vauquelin and Fourcroy, all themselves possessing chemicals manufactures), explained: 'Among all the factories . . . the alkali works have aroused strong demands which, unfortunately, are only too well-founded.'[41]

Montalivet then commissioned Costaz, the director of the Bureau des manufactures, who had made the case for the salt tax exemption, to draw up a draft decree. The minister already had a clear idea of the regulations he wanted to introduce, regulations that were both administrative and liberal, based on authorisation by prefects and reliance on civil justice. In April 1810, he recommended that the prefect of Hérault authorise a soda factory and refer 'to the courts any complaints to which the establishment may give rise'.[42] This, he explained, was the substance of the draft decree then being examined by the Conseil d'État. The prefect of the Bouches-du Rhône received the same instructions.[43]

The decree, published on 15 October 1810, confirmed these provisions. The government was responsible for issuing authorisation to factories, and the courts for arbitrating damages. Most power was concentrated in the hands of the prefect. Although the decree maintained the convenience inquiry as a prerequisite for authorisation, this was no longer an exercise in representing local notables, but a simple administrative consultation procedure. Compared with the inquiries that had been mounted under the Ancien Régime, nineteenth-century inquiries featured a more limited scope. They no longer reflected on what were the most appropriate economic activities in a given location, but merely listed the recriminations of proprietors defending the value of their properties. Finally, the result of the inquiry was judged by the Conseil de préfecture – that is, by four or five individuals appointed by the interior minister (and in fact by the prefect) – and not by the Conseil général which represented local notables.

The authorisation procedures were the core of the decree; after all, it was they that guaranteed industrial establishments their existence, regardless of any subsequent complaints the neighbours may make. First-class establishments (the most harmful, such as alkali factories) had to be located at a distance from residential areas. They were authorised by the minister, by means of a decree issued by the Conseil d'État, after a *commodo incommodo* inquiry in all municipalities located within five kilometres. Second-class establishments could be located close to

---

41   *Procès-verbaux de l'Académie des sciences*, vol. 4, 30 October 1809.
42   AD Hérault, 5 M 1014, Montalivet à Nogaret, April 1810.
43   Ibid., 3 May 1810.

residential areas and were authorised by the prefect after a *commodo incommodo* inquiry. Third-class establishments had to be authorised by the sub-prefects, who simply took the mayors' advice.

The classification does not seem to have caused any problems, except for the sulphuric acid manufacturers. The original copy of the 30 October 1809 report show that Deyeux had started by placing them in the first class (he had even started this column with them) before crossing them out and classifying them in the second.[44] In return, he added a note after the definition of the second class to the effect that the factories that could remain situated close to residential areas: 'It would only be desirable that the large mineral acid factories were always located at the outer limits of towns, in sparsely populated areas.'[45] In the end, the Conseil d'État ruled against the Académie and placed sulphuric acid factories in the first class. However, Article 11 was careful to specify that the decree did not apply retroactively – 'all active establishments will continue to be utilised' – and Article 12, which contradicted the previous one, added: 'However, in the event of serious inconvenience to public health, cultivation, or the general interest, first-class factories may be proscribed by virtue of a decree issued by our Conseil d'État.'[46] All this back and forth was a result of the sensitive case of the Chaptal factory in Les Ternes. When the decree was discussed in the Conseil d'État, it was difficult to forget that Chaptal, himself a member of this council, owned the largest sulphuric acid factory in Paris, in the *faubourgs*, and that he was also an associate of Holker, who owned the largest factory in Rouen.

These parleys over one or two factories produced a strange situation: the acid factories that had taken root in in the towns during the Revolution, and which were the source of the protests and the decree, had finally been legitimised! Better still, given the non-retroactivity of the decree, the first entrepreneurs took advantage of their locations on the edge of industrial cities, close to their markets and their workforce. They were also spared some expense on delicately transporting the heavy bottles of sulphuric acid. These commercial advantages explain the persistence of chemical industrial sites in town and city centres up till the end of the nineteenth century. Contrary to the classic justification for the 1810 decree, which explained that the variety of local policies had distorted competition, the new rules instead kept in place the distortions in the market.

---

44   AAS, session folders, 30 October 1809.
45   *Procès-verbaux de l'Académie des sciences*, vol. 4, 30 October 1809, p. 272.
46   Decree of 15 October 1810.

5. The right to pollute

The end of the environmental police and the division of work between the administration and civil justice system led to a redefinition of what was punishable. Under the Ancien Régime, environmental regulation relied on fines issued by the police, and pollution was punishable by law because of the damage it did to the common good. But thanks to the decree of October 1810, the major polluting industries were removed from the criminal law.

Let us return to the case of kelp. In the eighteenth century, the communes were granted a substantial arsenal of repressive measures: anyone who harvested kelp without having the right to do so, or outside the authorised period, or at night, or who uprooted kelp instead of cutting it, was liable for a fine of 300 livres – with corporal punishment for repeat offenders. The same penalties applied to those who burnt kelp during the summer or when there was a sea breeze. The fishermen's syndics were responsible for monitoring contraventions of the rules. An employee of the Admiralty emphasised the thoroughness with which they performed their duties: 'They draw up official reports, issue reprimands and turn off stoves lit in contrary winds. Convictions and fines are handed down immediately.'[47] Conversely, after 1810, the owner of a duly authorised chemical factory, who could destroy a river's fish stocks in the space of a few years by discharging industrial waste into it, did not run any risk of criminal penalty.[48]

The 1810 decree removed environmental damage from the remit of criminal law. This really shocked the neighbours of chemical plants who saw their crops destroyed: 'An unfortunate person whose devastation consists of a few grapes is punished as a devastator of rural property, yet [the industrialist], because he devastates, suffocates and suffocates on a large scale, may do so with impunity?'[49] Making industrial pollution a matter for the civil courts and not the criminal justice system raised fundamental problems: 'When the damage is involuntary, reason and the law only require compensation to be paid to the person who suffered it; but when this damage is repeated at every moment, day and night, and when the person who causes it intends to continue, I say that the matter degenerates into an offence.'[50]

---

47   AD Seine-Maritime, 1 B 5504.
48   Massard-Guilbaud, *Histoire de la pollution industrielle*, ch. 4, case 4.
49   AM Marseille, 23 F 1, *Mémoire pour Pierre-Joachim Duroure*, 1816, p. 8.
50   Ibid., p. 15.

The decree of October 1810 takes on its full historical meaning when we consider a text that immediately preceded it: the Penal Code, which had come into force that February. Article 471 of this code adopted the pre-revolutionary environmental penal regime. It covered a wide range of offences, including failure to maintain buildings or furnaces, obstruction of the public highway, out-of-control livestock and illicit gleaning. The article also included a clause condemning 'those who have thrown or exposed in front of their buildings, things of such a character as to cause damage by falling or through insalubrious exhalations'.[51] This provision could well have applied to the chemical industry, which was both booming and a focus of heated dispute at the moment the Penal Code was published. But it was circumvented by article 11 of the 1810 decree, which specified that the damage caused by factories would be addressed by the civil courts only. The 1810 decree was part of the restructuring of the economy of illegality linked to the development of capitalist society, as analysed by Michel Foucault: the Ancien Régime's abuse of rights (illicit gleaning for instance) became property infringement or theft. But at the same time, in the post-revolutionary recoding of the lawful and the unlawful, industrialists created a new legal order which made convenient exceptions for themselves.[52]

6. 'Hygienism', a pro-industrial agenda

The first hygienism (before around 1850) played a fundamental role in imposing industrialisation, despite the train of pollution that it brought with it.[53] Chemistry was its social and theoretical matrix. Of the four founding members of the Paris Conseil de salubrité, three were chemists: Antoine Parmentier, Deyeux and Cadet de Gassicourt. All were involved

---

51   Article 471 of the Criminal Code. The wording is taken from article 605 of the law of 3 brumaire an IV, 'Des peines de simple police'.

52   Foucault, *Discipline and Punish*, p. xxx. It should be noted that an 1859 ruling once again penalised industrialists for causing harm to watercourses. See Laurence Lestel, 'Le regard de la population sur l'industrie chimique, XVIIIe–XXe siècles', *L'Actualité chimique*, 355, 2011, p. 4.

53   Alain Corbin describes hygienism in this era as a 'cunning propaedeutic of progress'. See 'L'opinion et la politique face aux nuisances industrielles dans la ville préhaussmannienne', in *Le Temps, le désir et l'horreur. Essais sur le XIXe siècle*, Paris: Flammarion, 1991, p. 194. Thomas Le Roux fully confirms this analysis in his detailed study of Paris's Conseil de salubrité. Geneviève Massard-Guilbaud (*Histoire de la pollution industrielle en France*) draws up a more positive assessment of the work of the provincial conseils de salubrité, but this applies mainly to the latter half of the nineteenth century.

in industrial initiatives. In 1794, Deyeux and Parmentier joined forces to produce soda ash in the Somme.[54] Cadet de Gassicourt owned a large pharmacy in Paris and set his sights on heavy chemicals.[55] Two other industrial chemists, Jean-Pierre Darcet and Charles Marc, subsequently joined the board. We have seen how, in 1809, Marc tried unsuccessfully to take advantage of the soda boom. As for Darcet, he was a central figure in industrial and chemical networks. After working with Leblanc at the Monnaie de Paris on the process for making artificial soda, he set up a sulphuric acid and soda company in 1804. He then joined Holker and Chaptal at the factory in Ternes. He also worked with the chemist Jacmart at the head of a soda and soap factory. Through the intermediary of his nephew, he set up a sulphuric acid refinery, which provoked numerous complaints on rue Chapon, in the heart of Paris's Right Bank. In 1819, he invested 20,000 francs in Chaptal's chemical products company in Plan d'Aren.[56]

For these chemists, hygienism was above all a business of industrial improvement. According to Marc, 'it is the Achilles spear that heals the wounds it causes'.[57] Their aim was to make de-pollution financially profitable. The good entrepreneur, by optimising flows of materials and values, reduced both his losses and his pollution. The great model to follow was a Parisian entrepreneur named Jean-Baptiste Payen, who transformed the bulky piles of bones that cluttered certain streets in the north of Paris into ammonia.[58] In his company Payen kept, in addition to money accounts, an accounting of the chemical contents of the materials which it consumed, produced or rejected, in order to monitor their potential value.[59]

---

54   AN F12 2244, Rapport de la commission sur la soudière artificielle proposé par les citoyens Parmentier et Deyeux, 1794.

55   'Sur quelques nouveaux proceeds anglais', BSEIN, 1802, vol. 1, pp. 74–6.

56   APP, RCS, 1 November 1813 and 18 July 1820. AN F12 6728, Société anonyme du plan d'Aren.

57   Charles Marc, 'Introduction', AHPML, 1, 1829, p. xvi.

58   It is worth noting that this recycling could itself be particularly polluting. While the ammonia workshops did dispose of animal carcasses, they produced a foul-smelling oil. To get rid of it, Payen tried throwing it into the Seine (the oil spread 30 kilometres downstream and bothered Napoleon at Saint-Cloud), burning it ('a sort of black snow' fell on Paris) and when he threw it into a cesspool, the neighbours' wells became infected. See Alexandre Parent-Duchâtelet, 'Des inconvénients que peuvent avoir dans quelques circonstances les huiles pyrogénées et le goudron provenant de la distillation de la houille', AHPML, 2, 1830, pp. 16–41.

59   Jean-Baptiste Payen, Essai sur la tenue des livres d'un manufacturier, Paris: Johanneau, 1817; Marc Nitkin, 'Jean-Baptiste Payen et l'ombre de E. T. Jones', Histoire et mesure, 11, 1996, pp. 119–37.

Hygienists thus studied the urban productive fabric in its entirety, as a system of exchanges of matter producing value at each stage of industrial transformation. Their aim was to establish new connections between the different branches of manufacturing and to teach industrialists how to integrate their activity into the urban metabolism. The recycling economy characteristic of early-nineteenth-century Parisian industry corresponded perfectly with the liberal project of the hygienists seeking to make industrial profit compatible with urban health.[60]

More prosaically, sanitation was also of direct benefit to them, as it was based on chemical products they manufactured: acids, bleach and lime chloride. For example, tallow smelters (who made candles from animal fat) belonged to the first class, the most polluting, of establishments. The Conseil de salubrité nevertheless authorised them to be set up near dwellings on the condition that they used sulphuric acid, a technique patented by Darcet, member of the Conseil. All trades that processed large quantities of organic matter (including glue manufacturers, gut-string makers, renderers, beet sugar refiners, starch manufacturers) were thus driven to use acids to speed up their operations, deodorise and sanitise them. This policy opened up considerable business outlets for mineral chemistry.[61]

In practice, the Conseil de salubrité appeared to be more an extension of industrial circles within the prefectoral administration than a control body. In this respect, it was rather similar to the Conseil général des manufactures or the Bureau consultatif des arts et manufactures, which were made up of entrepreneurs charged with advising and lobbying the minister of the interior on their own affairs. The post-revolutionary regulation of industry that replaced the old guilds was a deceptive reform, for the administration set up a new system of self-regulation, on a national scale, run by the wealthiest entrepreneurs. In the first half of the nineteenth century, the pro-industrialist bias of the hygienists led to very high authorisation rates. Between 1811 and 1835, the Seine Conseil de salubrité authorised four-fifths of first-class establishments. Out of twenty-two

---

60   It is worth noting that this recycling could itself be particularly polluting. While the ammonia workshops did dispose of animal carcasses, they produced a foul-smelling oil. To get rid of it, Payen tried throwing it into the Seine (the oil spread 30 kilometres downstream and bothered Napoleon at Saint-Cloud), burning it ('a sort of black snow' fell on Paris) and when he threw it into a cesspool, the neighbours' wells became infected. See Alexandre Parent-Duchâtelet, 'Des inconvénients que peuvent avoir dans quelques circonstances les huiles pyrogénées et le goudron provenant de la distillation de la houille', *AHPML*, 2, 1830, pp. 16–41.

61   Le Roux, *Le Laboratoire des pollutions industrielles*, pp. 337–57.

applications for chemical plants, only one was rejected.[62] Between 1816 and 1850, the Bouches-du-Rhône prefecture authorised the establishment of seventeen acid plants and nine soda factories, and refused authorisation to only three projects (all in Marseille). In the *département* of Seine-Inférieure (Rouen), between 1818 and 1850, 777 out of 850 applications (91 per cent) were accepted.[63]

For opponents, this pseudo-hygienism and pseudo-regulation – which legitimised industrialisation in the name of science, and along with it the new power of the administration – represented a denial of justice. City dwellers were incensed by the collusion between industrialists and the experts who were supposed to regulate them. In Rouen, they denounced the two experts named by the prefect as 'personally interested in maintaining the acid factories'.[64] The inhabitants of Les Ternes humorously exposed Chaptal's blatant conflicts of interest:

> Their complaints are brought before the interior minister, who is this minister? M. Chaptal. The complaints were renewed . . . the new minister consulted the Institut, and the Institut adopted a report declaring that an acid factory was not unhealthy. Who was the author of this report? M. Count Chaptal. The local police are consulted. Who is the local police magistrate? Again M. Count Chaptal or, what amounts to the same thing, M. the Baron, his son. So, in this case, the offender, the expert, the senior judge and the inspector are all one and the same person.[65]

According to another opponent of a chemical plant, the minutes of the Conseil de salubrité 'are infected with fraud, forgery, absurdity, lies and charlatanism'.[66]

7. From the environmental to the social: the hygienic reconfiguration of aetiologies

Hygienists did more than simply propose technical solutions to environmental disputes. They also helped to redefine what the environment

---

62 *Archives statistiques du ministère des travaux publics, de l'agriculture et du commerce*, 1837, vol. 1, pp. 240–3.

63 AD Seine-Maritime, 5 M 318.

64 Ibid., 5 M 763, *Mémoire des habitants du Faubourg Saint-Sever*, 30 July 1806.

65 *Mémoire au Conseil d'État pour les habitants des Ternes contre M. le sénateur Chaptal*, 1811, p. 62.

66 *Conclusions de M. Bourgain, substitute du procureur du roi dans l'affaire de M. Lebel contre Paris et Graindorge*, hearing of 18 August 1827, p. 18.

was, or, more accurately, what it *could do*. In order to establish a financial system for compensating for environmental damage, it was still necessary to bypass a medical doctrine that held the environment as the determinant cause of health and disease.

The decree of 1810 distinguished between three categories of workshops: dangerous ones, where there was a risk of fire or explosion; insalubrious ones, where health was at stake; and the inconvenient, where the issue was merely the olfactory or auditory comfort of the neighbours. In the administrative vocabulary of the early nineteenth century, the notion of 'inconveniences' did not refer to the opposite of comfort, but to a situation that fell short of (and denied the real existence of) insalubrity. Defining the boundary between the inconvenient and the insalubrious was crucial, because it made it possible to make local residents' complaints toothless: while anyone could opine on what they found inconvenient, only the authorities and their hygiene experts could define what was unhealthy. Inconvenience related to the complainant alone, whereas unhealthiness is an objective property of the spaces studied by the science of hygiene.

In the early nineteenth century, the neo-Hippocratic framework remained dominant in medicine. Environment was the key to understanding health and disease or the cause of epidemics. In 1805, the great physician Cabanis explained that the aim of 'medical climatology' was to study 'the physical analogy between man and the objects that surround him'.[67] In Rouen, doctors were at the forefront of the fight against chemical factories. A petition signed by forty-two doctors was sent to the prefect warning him of the dangers of their fumes.[68] In Marseille, Doctor Fodéré openly criticised the soda makers and rejected the 1810 decree: 'In what code of the most barbaric nation is it written that the most natural right, the enjoyment of pure air, can be taken away?'[69] In Paris in 1834, the medical elite, faculty professors and leading clinicians opposed a 'salubrious knackering' project (that is, chemical products) supported by the Conseil de salubrité.[70]

The foundation of the *Annales d'hygiène publique et de médecine légale* by the members of the Paris Conseil de salubrité in 1829 emerged from

---

67   Pierre Jean Georges Cabanis, *Rapport du physique et du moral de l'homme*, Paris: Crapart, Caille et Ravier, 1805, p. 135.

68   Henri Pillore, *Quelques Mots sur les dangers des fabriques nouvelles d'acide sulfurique et de soudes factices*, Rouen, 1805; AD Seine-Maritime, 5 M 763.

69   Fodéré, *Traité de médecine légale*, vol. 6, p. 302.

70   Alexandre Parent-Duchâtelet, 'Des obstacles que les préjugés médicaux apportent dans quelques circonstances à l'assainissement des villes et à l'établissement de certaines manufactures', *AHPML*, 13, 1835, p. 286.

this context, in order to refute neo-Hippocratic concerns about the environment. Their explicit aim was to create a new medical speciality, to claim a monopoly on defining environmental risks and to reconfigure the way they were perceived by society. According to Alexandre Parent-Duchâtelet, a member of the Conseil de salubrité, hygienism should be seen 'above all in terms of its moral action on the minds of individuals and its influence on public opinion'; it should 'prove that in many areas, establishments that are inconvenient are not in fact harmful'.[71]

In order to refute environmental aetiologies, hygienists turned their fire on eighteenth-century studies of illnesses among tradesmen. For those working by the neo-Hippocratic paradigm, craftsmen had been fascinating objects of study: the substances they worked with and the vapours that surrounded them created a multitude of widely varying artificial microclimates, the comparative study of which was intended to elucidate the causes of disease. In 1776, the Royal Society of Medicine asked its correspondents to study 'the instruments used by workers, the materials they use, the vapours they produce . . . and whether these processes have had any influence on the prevailing epidemics'.[72] A long tradition, inherited from the *Morbus artificium* by the Italian physician Ramazzini (1699), took it for granted that the craftsman's body was moulded by his workshop environment. As late as 1822, the physician Pâtissier updated Ramazzini's treatise to include in his neo-Hippocratic approach the new trades that emerged with the Industrial Revolution.[73]

The first articles in the *Annales* concerning occupational hygiene may strike a rather surprising note: rather than focusing on unhealthy factories, they studied the workers' good health! The aim was to demonstrate to residents that the factories caused no harm. Parent-Duchâtelet and

---

71   Alexandre Parent-Duchâtelet, 'Quelques considérations sur le Conseil de salubrité', *AHPML*, 9, 1833, p. 250. On this strategy, which was part of the movement towards medical specialisation in the 1820s, see George Weisz, *Divide and Conquer: A Comparative History of Medical Specialization*, Oxford: Oxford University Press, 2005.

72   Meeting of 17 December 1776, quoted in *Journal de Paris*, 295, 22 October 1778.

73   Bernard-Pierre Lécuyer, 'Les maladies professionnelles dans les *Annales d'hygiène publique et de médecine légale*, ou une première approche de l'usure au travail', *Le Mouvement social*, 124, 1983, pp. 46–69; Ann La Berge, *Mission and Method: The Early Nineteenth-Century French Public Health Movement*, Cambridge: Cambridge University Press, 1992, pp. 148–83; Julien Vincent, 'Bernardino Ramazzini, historien des maladies humaines et médecin de la société civile?', in Julien Vincent and Christophe Charle (eds), *Contours de la société civile. La concurrence des savoirs en France et en Grande-Bretagne, 1780–1914*, Rennes: Presses Universitaires de Rennes, 2011; Thomas Le Roux, 'L'effacement du corps de l'ouvrier. La santé au travail lors de la première industrialisation de Paris (1770–1840)', *Le Mouvement social*, 234, 2011, pp. 103–19.

Darcet thus explained that they should study 'with the same care the professions whose effect is nil and even give them special attention [because] we are obliged to gather more facts to demonstrate the harmlessness of a factory than to prove its disadvantages . . . it is by these means that we do the greatest service to many manufacturers who practice their industry inside Paris'.[74]

Seeking to break any connection between location and health, hygienists compared the risks between different districts or between different professions. For example, by studying mortality rates, Parent-Duchâtelet demonstrated that the stinking environments of Montfaucon (a garbage dump in northern Paris) and the Bièvre (a small river sacrificed to industrial activity in the south) were not particularly unhealthy.[75] Statistics on professions were also abundantly used. For instance, workshop dust did not increase the risk of pulmonary tuberculosis because plasterers had less chance of dying of tuberculosis compared to writers. Similarly, contrary to the prejudices of eighteenth-century doctors, acidic atmospheres *reduced* the risk of tuberculosis.[76] Medical topography, that is the description of places in connection to health conditions, which had been a major genre in late-eighteenth-century medicine, was giving way to a statistical description of the health of the people according to their professions and income. Thanks to the medical surveillance of workers in certain large factories (the tobacco industry was a textbook case), hygienists had statistical sources (records of workers' sick days, for example) that enabled them to rule out environmental causes.

The first articles in the *Annales d'hygiène* thus transformed tradesmen's illnesses into woes resulting from some moral or material shortcoming. The illnesses of Parisian dockers did not owe to the insalubrity of the banks of the Seine, but to 'their habits and way of life'.[77] Benoiston pointed out that women were at greater risk of developing tuberculosis because of their low incomes, 'hence a state of discomfort that produces destitution and disease'.

---

74   Alexandre Parent-Duchâtelet and Jean-Pierre Darcet, 'Mémoire sur les véritables influences que le tabac peut avoir sur la santé des ouvriers', *AHPML*, 1, 1829, pp. 171–3.

75   Alexandre Parent-Duchâtelet, 'Recherches et considérations sur la rivière de la Bièvre', *Hygiène publique*, Paris: Baillière, 1836, p. 129.

76   While the risk for all professions combined was 114 per 1,000, for professions using acids (hatters or gilders) it was 76 per 1,000. Louis-François Benoiston de Chateauneuf, 'Influence des professions sur le développement de la phtisie', *AHPML*, 6, 1831, pp. 5–60.

77   Alexandre Parent-Duchâtelet, 'Mémoire sur les débardeurs de la ville de Paris', *AHPML*, 2, 1830, p. 265.

The work of Louis-René Villermé is a milestone in the history of medicine, famous for demonstrating the importance of living conditions and wealth as the main cause of life expectancy. Villermé is also famous for having inspired Engels's *Condition of the Working Class in England*. But in the 1830s this attention to social conditions was in fact the child of the hygienist and industrialist milieu studied above and in which Villermé operated. In 1831, he joined the Conseil de salubrité de Paris and was actively involved in refuting residents' complaints and the environmental aetiologies on which they were based. His seminal article of 1830, which linked the mortality of Parisian neighbourhoods not to the environment (such as the narrowness of the streets or the presence of workshops), but rather to the wealth of the people who lived there, participated in the programme of the founding generation of the Conseil de salubrité: using statistics to dismiss the sanitary effects of industrial pollution.[78]

More generally, professional diseases posed a problem to the liberal political economy of the early nineteenth century, for they directly incriminated economic activity. It was thus necessary to take the variety of artisanal climates that had fascinated eighteenth-century physicians and reduce them to the sole metric that economists could think about, namely wages. Adam Smith had paved the way. In *The Wealth of Nations*, after quoting Ramazzini, he utterly distorted the Italian author's argument: the illnesses of craftsmen owed not to the working environment, but to *excessive* work. Workers worked themselves to death because, lured by high wages, they worked too hard. And so, according to Smith, the solution was that wages should be limited![79]

Villermé's social hygiene played a similar, albeit somehow inverted theoretical role: it was not work itself that caused workers' suffering, but rather their low incomes. In 1840, in his *Tableau de l'*état physique et moral des ouvriers, Villermé was no longer worried about industrial environments:

Workshops are not exposed to these alleged causes of unhealthiness. They are singularly misunderstood when they are attributed illnesses that are mainly caused by forced labour, the lack of rest, the absence of care,

---

78   Louis-René Villermé, 'De la mortalité dans les divers quartiers de la ville de Paris', *AHPML*, 3, 1830, pp. 294–339. On Villermé and the creation of social hygiene, see William Coleman, *Death Is a Social Disease: Public Health and Political Economy in Early Industrial France*, Madison, WI: University of Wisconsin Press, 1982.

79   Adam Smith, *The Wealth of Nations*, London: Strahan, 1776, vol. 1, ch. 8.

insufficient food, habits of improvidence, drunkenness, debauchery and, to put it in a word, wages below real needs.[80]

Reducing tradesmen's health woes to a moral and financial issue justified a tempered economic liberalism. Industrialisation, whose very principles were challenged by residents' environmental complaints, became an acceptable historical transformation. Only a few adjustments would have to be made: controlling the bad habits of workers; an increase in wages to the level of 'real needs'; the abolition of child labour; and provident funds. Hygienism defined the biopolitics of liberal capitalism, that is, the minimal social conditions required to maintain the human labour force necessary for industry.[81] After demonstrating the correlation between human stature and wealth, Villermé explained that governments could 'improve the species at will . . . by working for the general happiness'.[82] It was the government's job to set in motion a virtuous circle, in which economic prosperity would create a stronger, more productive people. Political economy had replaced the environment (circumfusa) as the means of biopolitics.

The shift from medical topography to hygienic investigation – in other words from environmental to social aetiologies – made it possible to link industry and the progress of human health. Against the urban bourgeoisie, offended by the nuisances caused by industrialisation, the hygienists administered the repeated proofs that factories, despite the inconveniences they brought, were not only *not* unhealthy, but would bring a prosperous society and a healthier population. Of course, the reconfiguration of aetiologies was neither immediate nor monolithic. Throughout the nineteenth century, local residents continued to bemoan the ills produced by the factories. On provincial health boards, doctors sometimes spoke out against the theories of their Parisian colleagues.[83] But the key point lies elsewhere: the authorities, who had the final say on the authorisation of classified establishments, now had at hand the medical

---

80   Louis-René Villermé, *Tableau de l'état physique et moral des ouvriers employés dans les manufactures de coton, de laine et de soie*, Paris: Renouard, 1840, vol. 2, p. 209.

81   Hence Villermé's ambivalence: on the one hand, he played a decisive role in drafting the 1841 law restricting child labour; on the other, after the bloody repression of 1848, he wrote *Des associations ouvrières* ('Workers' associations'), one of the 'little treatises' commissioned by Cavaignac and intended to justify to the working classes the economic liberalism that had triumphed by force of arms.

82   Louis-René Villermé, 'Sur la taille de l'homme', *AHPML*, 1, 1829, pp. 388–91.

83   Fodéré, in particular, was critical of hygienism's methods, but also emphasised how isolated he was in this regard. See Fodéré, *Traité de médecine légale*, vol. 1, p. ix and vol. 6, p. 303.

theories and many pieces of evidence they needed to override all reference to man's environment. When, in the mid-nineteenth century, a dictionary defined a 'factory' as a 'dangerous neighbour', it appeared in Flaubert's *Dictionnaire des idées reçues* – a dictionary mocking received wisdom. Thanks to hygienism, industrialism had conquered the environment.[84]

8. How much is the environment worth?

Financial compensation for environmental damage was an absolutely generalised phenomenon. Case law shows that it concerned all types of business activities across France, throughout the nineteenth century.[85] What is more, civil suits account for only a minority of cases, as most financial transactions were conducted by mutual agreement. Chemical plants were particularly affected. The entrepreneurs generally preferred to compensate rather than stop polluting. Over the course of the century, a complex and changing body of case law developed on the subject: should compensation be paid for the depreciation of property values, for non-material moral damage, or only for material losses? Was it necessary to compensate for risk, future or potential damage (taking into account the increase in insurance premiums linked to a dangerous neighbour) or just the damage that had already happened? These legal decisions were critically important, since they defined the environment's monetary value.

This form of regulation through arrangements or civil judgments was the fundamental objective of the 1810 decree. The soda crisis provided the decisive turning point. In 1809, the Conseil de salubrité congratulated Barrera and Darcet for 'always being prepared to compensate farmers

---

84 French hygienism can be contrasted with British public health. Villermé's emphasis on social conditions as the main aetiological factor contrasts with the contemporary miasmatic theories of Edwin Chadwick. This was because the political contexts governing the redefinition of aetiologies differed. Chadwick's work aimed to justify the reform of the Poor Laws of 1834: it was necessary to show that mortality was not essentially linked to poverty or famine, but to dirtiness, such that sanitary techniques were considered to take precedence over social policies. French hygienists, unlike the English doctors, took an administrative role in authorising factories. Their desire to impose factories on city dwellers required them to construct an anti-Hippocratic aetiological framework that had points of contact with the social question of the 1840s. See Christopher Hamlin, *Public Health and Social Justice in the Age of Chadwick: Britain, 1800–1854*, Cambridge: Cambridge University Press, 1998.

85 See *Répertoire général alphabétique du droit français*, Paris: Sirey, 1900, vol. 20, pp. 801–7.

whose crops were damaged'.[86] That same year, Chaptal, who had just set up an alkali work on the Plan d'Aren near Marseille, signed notarised contracts to pay annuities to the neighbouring farmers for ten years. In Montpellier, the prefect noted with satisfaction that the operation of a chemical factory was 'continuing peaceably', thanks to the indemnities paid by the entrepreneur. Compensation for damage, first established as a quick fix, was then rationalised within a techno-liberal framework. By setting a price on pollution, it was supposed to produce the financial incentives that would spur entrepreneurs to innovate and reduce their emissions. For example, in 1823, the alkali manufacturers in Septèmes, near Marseille, complained to the government that the local civil courts were threatening the viability of their factories through the payouts they were awarding. The response from the Bureau consultatif des arts et manufactures was crucial: 'Justice must be allowed to follow its course, since the manufacturer condemned to very high compensation will soon be losing money, and will be forced to look for ways to condense the vapours... in this way, everything is respected and in accordance with existing laws.'[87]

The case of the alkali industry around Marseilles is decisive for the study of compensation. This is because, unlike most environmental conflicts during the early Industrial Revolution, it prompted a large-scale opposition, structured by the justice system.[88]

The alkali works were scattered across a rural area linked to the Marseille bourgeoisie: the olive groves provided oil for the soap-makers, the winegrowers sold their produce in Marseille and the wealthy city-dwellers owned country houses there. Opponents included small farmers, peers (d'Albertas, Simiane, Bourguignon de Fabregoules), merchants, doctors and prestigious lawyers (Louis Cappeau, the president of the Royal Court of Aix among others). This group made up the social and political elite of Provence. They were members of the Chamber of Commerce, the local councils of Marseille and Aix, and the General Council for the Bouches-du-Rhône. These institutions systematically advocated against the soda manufacturers. Unlike in socially segregated industrial towns – where pollution mainly affected workers who were integrated into paternalistic systems that did not challenge the environmental order – the heterogeneous social fabric around Marseille explains the strength of the mobilisation there.

86   APP, RCS, 4 May 1810.
87   AN F12 4783, 18 January 1823.
88   Xavier Daumalin, 'Industrie et environnement en Provence sous l'Empire et la Restauration', *Rives nord-méditerranéennes*, 28, 2006, pp. 27–46.

This was especially true insofar as the alkali manufacturers were often strangers to Provence: Mallez came from Valenciennes, Kestner from Strasbourg, Dubuc from Rouen, Bonardel from Lyon, and Pluvinet and Chaptal from Paris. At the beginning of the Restoration, their position appeared greatly weakened: the port exemption granted by Louis XVIII would restore trade in natural soda from Spain and Egypt, and it seemed obvious that the harmful factories supported by the fallen regime would be banned. In 1815, a farmer jeered at a soda manufacturer whom he came across in Marseille: as soon as the king was well in place, he would surely order the destruction of the factories and throw the manufacturers in jail. Two events prompted the disappointment of such expectations: in 1815, an ordinance authorised the import of natural soda ash, but with a prohibitive duty of 15 francs per quintal, and in May 1816 the prefect announced a *commodo incommodo* inquiry in order to authorise a new factory in Septèmes. Furious, the residents went en masse to the prefecture. The lawyer Romieux harangued them: 'It is all the same! If the Prefect does not give you good reasons, you will take the law into your own hands!' On 4 August 1816, a letter from the interior minister was posted in Marseille and Septèmes: it stated that the prefect was the guarantor of the manufactories' existence, and would use force if necessary.[89] The post-Restoration government publicised its continuation of the imperial policy. It guaranteed the continued existence of the factories despite the general opposition of both the rural population and the notables of Marseille and Aix-en-Provence.

The central government rejected complaints from residents, but also from the local and *département*-level authorities (Marseille municipal council, and the General council of the Bouches-du-Rhône). Accordingly, the fight against the factories moved into the legal arena. From 1816, ever more suits demanding compensation were brought before the justice.[90] At first, the manufacturers took the matter lightly. Mallez appointed a proxy to handle 'this little legal war'.[91] But, from 1817, they challenged the decisions of the first level of jurisdiction (juge de paix) and decided to systematically appeal their rulings. Their aim was to discourage small farmers who could not afford the advance payments to bring a trial to the civil court. This strategy initially seemed to bear fruit: in 1817 and

---

89   AD Aix-en-Provence, 208 U 19, 1818 and CCI Marseille, *Mémoire pour les sieurs Mallez Frères, fabricants de produits chimiques à Septèmes, contre les opposants*, 1818.

90   AD Bouches-du-Rhône, 4U 12 6 to 4U 12 9 (justice de paix du canton de Gardanne).

91   Ibid., 4U 12 9.

1818, dozens of out-of-court settlements were reached. But in hindsight, it was a big mistake.

Some sixty farmers continued their legal battle. Their surprising perseverance was made possible by lawyers who decided to take up the fight against the soda factories. Seytres and Romieux were particularly active. According to a report by the public prosecutor, they sent emissaries to the countryside to persuade the peasants to sue the manufacturers, and they advanced the legal costs, promising to take their payment from the compensation that resulted. In 1823, they reportedly earned the considerable sum of 50,000 francs from industrial pollution lawsuits.[92]

For legal experts, the industrial pollution cases were above all a conflict of jurisdiction between the Provençal courts and the centralised system of administrative litigation established by Napoleon. In 1815, the magistrates of the Royal Court of Aix, many of them veterans of the *parlement* of Provence, tried to defend their autonomy against the Court of Cassation and the Council of State. To their eyes, it seemed odd, to say the least, that a judicial decision could not overturn the authorisation given by a prefect. After all, the latter had replaced the police measures of the Ancien Régime, the lowest-ranking texts in the hierarchy of norms. From an eighteenth-century perspective, this was tantamount to saying that a police order took precedence over a *parlement*'s ruling. The prefect's decision did not respect any judicial form; it was based on investigations and reports that were neither based on an adversarial procedure, nor even public. Lastly, the prefect's authorisation for a factory 'at most only forms a *presumption* that the neighbourhood will not suffer from this establishment'.[93] This was merely an expectation, which could turn out to be wrong. The real existence of damage – as duly established by expert reports and confirmed by a court of law – naturally ought to nullify an erroneous administrative measure. It was simply absurd to prohibit the courts from challenging an administrative act. In the first half of the nineteenth century, the separation of powers between the administration and justice meant above all the reduction of local judicial power and the notables who controlled it. To put it differently, a judicial court's assessment of the present could not refute an administrative authorisation based on forecasts. This paradox, which was part and parcel of the administrative revolution of the early nineteenth century, made it possible to protect investments from the

---

92   CCI Marseille, D 1520, 'Rapport du procureur général au garde des Sceaux, 28 juin 1824'.

93   Louis Cappeau, *Traité de législation rurale et forestière*, Marseille: Ricard, 1823, vol. 1, p. 396.

vagaries of the law and to force through the establishment of new factories, even against the wishes of local notables.

The legislator had so restricted the power of the judiciary, now able
neither to ban a factory nor to deem it unhealthy, that the Marseille civil
court and the Royal Court in Aix had to innovate. Deprived of the power
to suppress licences to factories, they made full use of their powers to
arbitrate disputes and award compensation. To make manufacturers pay
the price of pollution, and perhaps even more than this, the Provençal
courts used the concept of 'moral damage'. This elastic category made it
possible to give legal existence to the environment, in the form of quality
of life: the beauty of a landscape, the purity of the air and water, and the
comfort of country life became monetisable.

Just a few lawsuits would cost the alkali entrepreneurs a lot of money.
Septèmes counted four large estates. In 1822, Fabregoule claimed 100,000
francs in damages. He was awarded 24,000 francs for loss in land value
and an annual annuity of 4,000 francs for reduced harvests. A few other
large landowners also received compensation in excess of 10,000 francs
each. But for small properties the material damage was necessarily limited:
a few dozen, sometimes a few hundred francs at most. These cases were
nonetheless ruinous for the manufacturers because of the costs they
generated in terms of seeking expert appraisals.

Evaluating the damage that acid vapours did to crops was a tricky
business. First, the causal link had to be established. The manufacturers
asked the experts to explore various alternative hypotheses: poor crop
maintenance, exhausted soil, plant diseases, the influence of a salty wind
coming from the sea (*pouvarel*) or even a general deterioration in the
Provençal climate.[94] Then, in order to determine the damages, it was
necessary to estimate the proportion that resulted from acid vapour
within this set of causes. The experts had to specify and refine the general
theories in order to apply them to the specific case: to this particular
property, situated on a certain type of soil and subject to the influence of
this or that wind. For example, in 1828, a farmer in Septèmes complained
that he could no longer use his washhouse because the contaminated
water no longer dissolved the soap.[95] The problem was a difficult one:

---

94   ASM, 'À monsieur le maire de Saint-Mitre', 1824. The summer of 1817 was
exceptionally dry, and in the winter of 1821 the olive trees froze. Ségaud, the director of
the Plan d'Aren factory, pointed out that the compensation claimed by his neighbours
varied according to these climatic accidents. He also mentioned the *pouvarel*, a sea wind
that deposits salt on crops. See also 410 U 32, Foucard v Bertrand, 1822.

95   AD Bouches-du-Rhône, 410 U 40, Antoine Poutet contre Quinon, Cusin,
Rougier, Grimes, 1828.

the water in the washhouse did contain lime, but lime is found in all water in limestone regions. The problem was thus quantitative, and 'no scientist has yet determined the maximum amount of these bodies'. The experts therefore analysed the water in the Marseille region to establish the *normal* concentrations of these salts.

This judicial production of knowledge on environmental damage – infinitely richer and more detailed than the administrative reports – was very expensive: between 200,000 and 300,000 francs for the hundred or so reports produced between 1819 and 1835.[96] Whatever the compensation awarded, the costs of the expert appraisals were passed on to the soda makers, who were ordered to pay them. For example, the assessment of the washhouse cost 2,993 francs to award annual damages of 40 francs, which compensated the farmer for using a neighbouring washhouse. The appraisal was skilfully manipulated by a small group of farmers and supported by the judges, who accepted their demands for extraordinarily rigorous and costly experiments. The experts had no qualms about writing outright dissertations in chemistry and botany (often running to over 100 pages) to assess the damage done to a few vines. At 20 francs a day, expert appraisal work was a godsend.

The third factor that made pollution costly was the rigorous application of article 1382 of the Civil Code: 'Any act of man that causes damage to another person obliges the person through whose fault it occurred to repair it.' The problem was the meaning to be given to the word *damage*. The civil court of Marseille and the Royal Court of Aix ruled that the soda makers had to compensate their neighbours three times over: for material damage (loss of crops); for immaterial damage (reduction in the market value of the property); and for 'impairment of enjoyment' (or non-material damage).

Moral damage was based on the loss of a way of life associated with owning a rural property: the *bastide*. For the Marseille merchants, these residences signalled their membership of the social elite and demonstrated their ability to manage an estate well. In 1816, an owner asked for compensation because he had to leave his estate, and giving up his inheritance was a *humiliation*. He also had to be compensated for the many inconveniences of living in the city (neighbours, noise, bad air).[97] The enjoyment of a beautiful rural landscape was also a factor: unobstructed views of the sea, Marseille or wooded valleys were valued. Conversely,

---

96   When the Académie de médecine was founded in 1823, it was given a budget of 40,000 francs a year.

97   AM Marseille, 23 F1, *Memorandum pour Pierre-Joachim Duroure*, 1816.

the requests of the Rouxs, who owned a 'calanque' (steep-walled inlets near Marseilles), were unfounded, for 'what pleasure c[ould] be offered by an arid property filled with precipices and hideous rocks'?[98]

The fight against the factories intersected with a movement which invented the Provençal tradition, after and in opposition to revolutionary and imperial centralisation. For instance, opponents included the founders of *La Ruche Provençale* (1819), a literary and scholarly journal featuring most of the themes of Provençal Romanticism (troubadours, shepherds, the good King René), as well as attacks on the soda makers.[99] The 1820s also saw the cultural development of the idea of Provence as a beautiful landscape. Septèmes, on the road from Aix to Marseille, was featured in travel guides because of its picturesque valleys and views over the harbour. A local aristocrat, Bourguignon de Fabregoules, was associated with the famous painters Jean-Antoine Constantin and François-Marius Granet, whose favourite subjects were the landscape around Marseilles and the *bastides*.[100]

By forcing the soda makers to compensate for subjectively felt damages, the civil courts were opening up a Pandora's box. No longer just the loss of crops, but all forms of environmental damage could now be taken into consideration. No matter what the health experts said, it was impossible to rein in the fears of those forced to bring up their families amid acid fumes. Opponents thus called for damages to compensate not for illness, but *the fear* of illness.[101] The order of magnitude of compensation changed: in addition to the crops, the soda farmers had to compensate for property depreciation (up to 30 per cent of its value) and pay considerable annuities for impaired enjoyment (5 to 10 per cent of its value per year).

This jurisprudence, which put the civil courts back at the heart of regulation, lasted only for a short period, between 1822 and 1827. In this last year, the lawyer Louis-Antoine Macarel defended the soda makers before the Cassation Court. He drew on Roman law to distinguish between material damage, or *damnum illatum*, and immaterial damage, *damnum infectum* (damage that has not yet occurred): 'The assessment of this damage cannot fall within the remit of the courts, as it prompts *preventive action* by the state's general police force.'[102] By hardening the distinction

---

98   AD Bouches-du-Rhône, 410 U 46, Roux contre Daniel et Cie.

99   Laurent Lautard, 'Sixième lettre sur Marseille', *La Ruche provençale*, no. 2, 1820.

100   Guy Cogeval and Marie-Paule Vial, *Sous Le Soleil, exactement. Le paysage en Provence du classicisme à la modernité (1750–1920)*, Ghent: Snoeck, 2005.

101   *Mémoire pour Pierre-Joachim Duroure.*

102   Louis-Antoine Macarel, *Requête, les sieurs Armand et Cie fabricants de soude artificielle au lieu de Couran, commune d'Auriol, département des Bouches-du-Rhône*, 1826.

between prevention and reparation, Macarel's doctrine rendered the civil courts powerless. Since the loss in value of a property is a *damnum infectum* (it may be realised at the time of sale), it fell outside of the jurisdiction of the civil courts. In several cases in both Paris and Marseille, the Cassation Court followed this principle.[103] Despite the opposite position of the lawyers defending the Royal Court of Aix's judgment, this precedent was immediately taken up by de Gérando in his *Institutes du droit administratif*, the first post-revolutionary manual of administrative law and in the first legal treatises on dangerous establishments.[104] In 1830, Macarel became a Conseiller d'État. He was one of the great theoreticians of French administrative law, and his *Cours d'administration*, which was constantly republished and updated, served as the basis for nineteenth-century administrative case law. The victory of administrative law over the civil law doctrine of damage was fundamental because it prevented compensation for losses in land value. It abruptly devalued the environment and completed the project of the 1810 decree. For now, the joint management of industry and the environment fell (almost exclusively) under the remit of the administration.[105]

9. Shaping technology, or being shaped by it

What were the effects of this commodification of the environment?

Technologically speaking, they were almost non-existent, and the state had to intervene. In 1823, the interior minister ordered the alkali industrialists to condense their hydrochloric vapours. They had one year to comply. If they failed, they would have to close their factories. The situation was truly critical, since all attempts to condense vapours had failed up to that point.[106] Rougier, a soda maker from Septèmes, experimented with a rustic, monumental and relatively effective system: he dug

---

103 Alphonse-Honoré Taillandier, *Traité de la législation concernant les manufactures et les ateliers dangereux, insalubres et incommodes*, Paris: Nève, 1827, pp. 46–56, 153.

104 Joseph-Marie de Gérando, *Institutes du droit administratif ou éléments du code administratif*, Paris: Nève, 1829, vol. 3, p. 308. Compare with Jean-François Fournel, *Traité du voisinage considéré dans l'ordre judiciaire*, Paris: B. Warée, 1827, vol. 2, p. 50.

105 It should be noted that in 1850, the Cour de cassation again ruled in favour of offsetting capital losses on land. See *Répertoire général alphabétique du droit français*, vol. 20, p. 802.

106 Clément-Desormes, a professor at Arts et métiers, and Péclet, a Parisian chemist, had been brought in by the Marseille soda manufacturers, albeit all in vain; see H. de Villeneuve, 'Les condensateurs des fabriques de soude', *Annales de l'industrie du Sud*, 2, 1832, pp. 129–44.

1-metre-deep and 600-metre-long trenches in the limestone hills above his factory, covered by a vault of limestone rubble. A stream of water and steam circulated inside. Basins collected the condensed hydrochloric acid.

In fact, the problem was both technical and political. Condensation was very expensive: the limestone conduits eroded and had to be rebuilt every year; they reduced production rates and cut yields by a third.[107] The Rougier apparatus could thus only be effective where, and so long as, landowners were prepared to take civil action. Contrary to technophile accounts of this case, it did not put an end to lawsuits but simply made them rarer.[108] While some manufacturers did indeed opt for condensation (in Septèmes), others, located in a less densely populated habitat or a less hostile environment, continued to pay compensation.[109] Technology made it possible to strike a fine balance between the cost of production and the damage caused; it was the site of compromise between manufacturers and landowners. We ought to think of this de-pollution as a dynamic process: it was both a question of devising the right technique and having the local mobilisation that forced the industrialists to use it.

If, in Septèmes, the mobilisation was sufficiently powerful to shape the technology, elsewhere – in the ponds near Istres, in Salindres, in Dieuze, in Thann, or in the British case, in Widnes and St Helens – it was the soda works that shaped environments and societies. Here, entrepreneurs protected themselves by creating small industrial colonies and taking control of the towns that depended on them.

Take the case of the Plan d'Aren company. Its location, in the middle of the ponds between Fos and Saint-Mitre, did not make strict condensation a compulsory choice, and the company opted to issue compensation rather than maintain a condenser. In a confidential note, the director explained: 'If it were necessary to talk about a condenser *to paralyse any recriminations*, we could make a simulacrum of a device that would be worth what it could do.'[110] The solution relied more on controlling the social environment around the factory. The company acquired meadows several kilometres away, which were otherwise of no interest, in order to create a patronage network. In 1821, one major farmer in Istres made a deal: he withdrew his complaint and, in exchange, the company granted

107   A condenser cost 15,000 francs to install. Repairs cost 6,000 francs a year.
108   Louis Figuier, *Les Merveilles de l'industrie*, Paris: Jouvet et Cie, 1873, vol. 1, p. 492.
109   AD Bouches-du-Rhône, 410 U 43, Daniel and Cie v Vagalier, 1828 and 410 U 46, Daniel and Cie v Roux, 1830.
110   ASM, Note sur l'enquête pour servir de renseignement, 7 March 1824.

him the right to use its considerable grazing land.[111] Other transactions were also possible: granting pasture rights, authorising a farmer to use a well or an oven, reducing the rent on the land in the event of a poor harvest, or hiring the children of farmers.[112] The company became the main employer in the region. In 1824, it declared that it employed 400 people and spent 300,000 francs a year on wages. At the time, Istres had a population of less than 3,000 and the olive oil trade in the canton was estimated at 100,000 francs. Many small-scale farmers, whose yields were falling due to pollution, were employed by the day in the salt marshes, while their children worked in the factory. In the 1850s, the Compagnie générale des produits chimiques du Midi, which had taken over the factory, opened a school and a church for its workers. A general store sold the workers goods at cost price. It was also open to villagers.[113]

The chemical industry was thoroughly transforming rural societies. In Istres, there were two opposing factions: the Fumado (smoke, in Provençal), standing for industrial interests, and the Plouvino (white frost), which represented agricultural interests. Each clan has its own church, school, dances and shops. Local political life was structured by the factories. In 1830, Jean Cappeau, a Plouvino landowner, lost the mayoralty to Auguste Prat whose son was director of the Compagnie générale des produits chimiques and would himself became mayor of Istres in 1854. When a complaint was sent to the Senate, calling for the factory to be shut down, the government refused to intervene on the grounds that it was a petty struggle between local notables.[114] So here, the environment informed the local political struggle, which in turn made environmental complaints inadmissible.

The environmental degradation and social processes that made it possible were similar in all the towns where chemical companies were founded. From the 1850s onwards, the railways made it possible to set up factories far from towns, and the sector became more concentrated: Salindres, Dieuze, Thann and Marseille were the home of almost all of France's alkali production. In England, production was concentrated in St Helens and Widnes, near Liverpool. Alakli works required the same

---

111   Ibid., Transaction entre Imbert et la Cie, 1821.

112   Ibid., Réponse de la Cie au maire de Saint-Mitre, 1824.

113   The forced exile of the factories led to the early development of paternalistic practices. Soda makers who moved to the islands of Porquerolles and Port-Cros were forced to house and feed their workers. See Xavier Daumalin, 'Patronage et paternalisme industriels en Provence au XIXe siècle: nouvelles perspectives', *Provence historique*, 55, 2005, pp. 123–44.

114   AN F12 4983, Petition to the Senate, 25 March 1868.

basic ingredients (a rail link, proximity to coal or salt) and produced the same effects: the creation of industrial colonies *ex nihilo*, in the middle of the countryside, the establishment of paternalistic structures, the conquest of local power by manufacturers – and massive environmental degradation.

When Charles Kestner set up his factory in Vieux-Thann in Alsace in 1807, the village had 500 inhabitants. The factory stagnated until the opening of the Thann–Mulhouse railway in 1846. In 1860, out of a population of 1,120, Kestner fils employed 330. Relief funds were established for workers and their families; the town hall and then the *département* came under the political control of this powerful family of industrialists.[115] In the 1820s, in Dieuze, Moselle, the manager of the soda works preferred to pay damages (5,000 francs a year) rather than condense the fumes. Baron de Prel, mayor of Dieuze and a member of the Conseil general of Moselle, was one of the main shareholders in the firm.[116] In 1845, an article in the *Annales d'hygiène* described a situation with nothing in common with the one which prevailed in the area around Marseille: 'torrents of vapour enveloped the town', 'the earth was bare and barren, the grass was burnt', the ironwork was corroded, and litmus paper turned red even at a kilometre's distance.[117] Faced with the flood of complaints, the prefect explained: 'We have to take care of an establishment that, so to speak, brings life to the town of Dieuze.'[118] Similarly, in the 1870s, the Péchiney soda factory employed almost half the population of Salindres. Henri Merle, the plant's director was, of course, also the town's mayor. Despite complaints from farmers, the factory was regarded by hygienists as a model establishment, thanks to the medical service and shops it provided for the villagers.[119]

In England, the Leblanc process for producing soda appeared later than in France, in 1823 when Muspratt and Gamble set up a factory north of Liverpool. Complaints flooded in immediately, forcing the manufacturers to leave the city and move to St Helens. For the proprietor, this small market town, located on the railway line between Manchester and

---

115   AD Haut-Rhin, 5 M 99.

116   *Mémoire contre l'établissement d'une fabrique de produits chimiques à Épinal*, 1830, p. 17.

117   Henri Braconnot and François Simonin, 'Note sur les émanations des fabriques de produits chimiques', *AHPML*, 40, 1848, pp. 128–37.

118   AN F12 4937, le préfet au minister de l'Agriculture, du commerce et des travaux publics, 17 October 1867.

119   Laurent Roch, *Rapport au conseil d'hygiène de l'arrondissement d'Alais*, Alès, 1880.

Liverpool, offered the advantage of being 'sufficiently large to house his workpeople yet sufficiently small not to possess any organised form of local government which might restrict the growth of his factory'.[120] The alkali industry also spread to Widnes, a hamlet near St Helens, which was exclusively agricultural in 1830. Twenty years later, these villages had become the world centre of alkali production: the fifteen factories amassed there produced 120,000 tonnes a year. According to many complementary eyewitness accounts, the damage was beyond comprehension. *The Times* described an apocalyptic landscape, with not a single living tree for miles around. In 1862, the House of Lords launched an enquiry. Residents testified to the environmental turmoil: the orchards and hedgerows of this bocage land had disappeared, making it impossible to raise livestock; mountains of waste filled the atmosphere; residues from alkali manufacture sometimes used as backfill for canals, oozed a yellowish liquid that killed fish. The waterways were so acidic that steamboat firms complained about the rapid corrosion of their hulls.[121]

In 1862, the House of Lords, which had taken up this issue, emphasised the ineffectiveness of existing pollution laws: the Public Health Act, the Local Improvement Act and the Smoke Prevention Act provided powers to local authorities, but the latter were not obliged to make use of them.[122] In small industrial centres such as Widnes and St Helens, the industrialists who held sway politically were thus protected. The other difficulty stemmed from the characteristics of the English courts. In France, the law allowed a private individual to sue all the alkali producers in a given territory, with damages being paid in proportion to their production.[123] In England, on the other hand, landowners had to prove the causal link between the damage and *one specific* factory. When such establishments were grouped together, as in St Helens and Widnes, and huge chimneys (some as much as 100 metres tall) were built, it was impossible to prove the causal link.[124]

---

120   Theodore Barker and John Harris, *A Merseyside Town in the Industrial Revolution: St Helens, 1750–1900*, Liverpool: Liverpool University Press, 1993, p. 228.

121   'Report from the select committee of the House of Lords on injury from noxious vapours', *House of Commons Papers, Reports of Committees*, 1862, vol. 14, question 164.

122   Eric Ashby and Mary Anderson, *The Politics of Clean Air*, Oxford: Clarendon Press, 1981; Peter Thorsheim, *Inventing Pollution: Coal, Smoke, and Culture in Britain since 1800*, Athens, OH: Ohio University Press, 2006, ch. 8.

123   Such was the case law adopted by the Provençal courts and confirmed by the Cassation Court. See Taillandier, *Traité de la législation concernant les manufactures*, pp. 167–9.

124   'Report from the select committee of the House of Lords on injury from noxious vapours', questions 164, 220, 887.

William Gossage, a manufacturer from Widnes, had developed a condensing technique but, in England as in France, its effectiveness depended on the social conditions in which it was used. The smaller industrialists rejected condensation, believing it to be a subterfuge that allowed manufacturers to impose their control over the market. Moreover, as the workers were paid by the tonne of soda produced and the condensing towers reduced the speed and yield of their work, they had no interest in using them. Particularly at night, they sent the vapours directly through the chimneys.[125]

10. Invisible workers

On 28 July 1893, in a Liverpool hotel room, a British Home Office committee held hearings with seven workers from the alkali works at Widnes and St Helens. Most of the workers had declined the invitation to give evidence, for fear that it would get them into trouble. The committee's chairman reassured them: 'The masters will know nothing about your evidence now. You may speak quite freely.'[126] They all described the stigma of their work: breathing difficulties, perforated septums, and the loss of smell and taste. Most of the workers in Widnes had lost their teeth because of the acid fumes. The manufacturers provided no protective equipment. Before entering the lead chambers, the workers placed a flannel cylinder in their mouths, covered their clothes with newspapers and smeared their faces with grease to protect themselves from the fumes.[127] Most wore homemade protective goggles, but this was not enough to prevent them frequently being splashed in the eyes with acid.

Here is the deposition of Robert Hankison, a worker from St Helens, who spent fifteen years working between pots of sulphuric acid. He was asked what the workers did when one of them got acid in their eye.

– I think the method you have there is getting a man to fill his mouth with water and squirt it into the eye?
– Yes sir, of course, but if there is not too much in you get him to put his tongue in your eye and lick it out.

---

125 Ibid., questions 1,656–9. Hydrochloric acid vapours circulated in a tower filled with coal and pulverised water.
126 'Report to Her Majesty's principal secretary of state for the Home Department on the conditions of the labour in chemical works', *British Parliamentary Papers*, 1893, question 468.
127 Ibid., question 492.

Alkali workers[128]

Are these splashings very frequent?
– No, sir, mine was quite an accident. It might not happen in a hundred years.
– When did you see a man get splashed last?
– About two days ago.
– How often will it occur?
– It just depends how neglectful they are.
– Once or twice a week?
– Oh no, sir, it was a matter of neglect that the man got it the other day. He said he would not do it any more, but it was too late then, he had got the splash.
– In regard to these pots, you have had further experience. I believe your father met with his death at these pots, did he not?
– Yes, sir.
– Can you tell us how that happened?
[Hankinson describes his father's imprudence] . . . Of course he would not have met with the accident if he had not had his pot so full as he had it.[129]

---

128   Robert H. Sherard, *The White Slaves of England*, London: Bowden, 1898.
129   'Report to Her Majesty's principal secretary', questions 155–81.

Two things are worth noting here. First, the constant downplaying of risks by the workers themselves (the accident meant to happen only once in a century had taken place just two days earlier). Second, the workers considered themselves responsible for accidents. Their injuries owed to negligence, inattentiveness, bluster, a desire to rush everything. One worker injured by an acid splash explained that he had been 'punished' by the tank of caustic soda.[130] In 1881, the Alkali Act, which imposed safety measures, prompted industrialists to issue factory regulations which workers could be fined for breaking. In 1893, workers complained because they were the ones who had to bear the additional costs of safety: the regulations, by forcing them to work more slowly, reduced their incomes. By paying piece-rate wages, entrepreneurs subjected workers to the capitalist logic of maximising output.

The near-total invisibility of this extraordinary suffering until the end of the nineteenth century was the result of specific means of managing the workforce: importing foreign workers, the medical selection of workers, and a rapid staff turnover.

For example, in the alkali works of Provence, dangerous jobs were handed to Italian workers – generally single men, whose illnesses and deaths would have little impact on local communities.[131] In the match industry, which was particularly dangerous because of the jaw necrosis caused by phosphorus, the workers were regularly inspected by dentists. If there was the slightest doubt, they were dismissed. Until the end of the nineteenth century, the aim of industrial hygiene was less to clean up the trades than to select workers suited to them.[132] In English factories, it was accepted that mortality statistics were meaningless because workers in poor health were dismissed before they died. A doctor confirms: 'I told [the management] privately . . . that the man would not live two years . . . they banished him out of the works altogether, and he died as a bricklayer's labourer.'[133]

In the ceruse (or white lead) factories, the workers all suffered from painful lead colic. Industrialists had no choice but to hire ex-convicts or

---

130   Ibid., question 428.

131   In 1820, thirty-four of the seventy-two workers at the Bérard soda factory were Genoese or Piedmontese; they were aged between eighteen and fifty-five, and only four were married. AD Bouches-du-Rhône, 116 E D4, État nominatif des ouvriers employés à la fabrique de soude indiquant leur lieu de naissance et leurs départements, 1820.

132   Caroline Moriceau, Les Douleurs de l'industrie. L'hygiénisme industriel en France, 1860-1914, Paris: EHESS, 2009, ch. 2.

133   'Report from the select committee of the House of Lords on injury from noxious vapours', question 598.

to organise a rapid turnover of their workforce. At the factory in Clichy, near Paris: 'Every fortnight there was a change of workers. They left the factory and went off to do other things, and in this way they avoided the illness that they would unfailingly have contracted had they stayed longer.'[134] The poet Charles Gille (1820–56), who had been a white-lead worker before becoming a lieutenant in the Republican Guard in 1848, expressed the resignation of this *Lumpenproletariat*:

> By fates unlike each others'
> To misfortune driven down
> At the cabaret, my brothers
> Our sorrows we can drown
> Without worry or remorse, let's get drunk, friends
> At work, tomorrow, we'll meet our deadly ends[135]

The chemical industry needed this destitution: finding a workforce willing to take such risks was not easy. As early as 1798, Chaptal explained that the liberalisation of the labour market through the abolition of guilds demanded a rebalancing, and that the chemical industry in particular required a form of constraint. Since the worker was 'too often disposed to refuse *difficult or disgusting* operations, a compulsive force is needed to *constrain* him to [do] it. However, this force exists only in the ties that bind him to the workshop and *place* him at the *disposal of the boss*.'[136]

Until the 1880s and the major public controversies surrounding the fate of matchmakers and white lead workers, French hygienists paid little attention to occupational health. For example, in the nineteenth century, the *Annales d'hygiène* published only one article devoted to acid and alkali factories. The deteriorating health of the workers was mentioned in its pages, as a rumour: 'We have had reports of workers losing their teeth, suffering from purulent ophthalmia and lung problems.'[137] The first occupational hygiene study on alkali works in France, conducted by a medical student from Nancy, dates to 1881.[138]

---

134   Brechot, 'Mémoire sur les accidents résultant de la fabrication de ceruse', *AHPML*, 12, 1834, pp. 72–5.

135   Edmond Thomas and Jean-Marie Petit, *Voix d'en bas: la poésie ouvrière du XIXe siècle*, Paris: Maspero, 1979.

136   Chaptal, *Essai sur le perfectionnement des arts chimiques*, pp. 9–10.

137   Henri Braconnot and François Simonin, 'Note sur les émanations'.

138   Albert Olivier, *Des Dangers ou inconvénients professionnels et publics de la fabrication de la soude*, Nancy, 1881.

Even among those doctors who stuck by the environmental medicine paradigm, there were efforts to downplay risks. The notions of habit and acclimatisation played an essential role, here: the workshop was conceived as a microclimate to which the worker must accustom himself. Dr François Mêlier offered a striking analogy:

> The position of a worker, approaching certain workshops for the first time, is somewhat comparable to that of a traveller who finds himself transported under a new and different sky; like him . . . he has to mould himself under the action of other elements; in a word, to undergo the tests and modifications of a kind of acclimatisation.[139]

Referring to a phosphorus factory, the hygienist Alphonse Dupasquier explained that, despite a harsh first impression, 'the workers quickly got used to it, *acclimatised to* it, and then lived in the midst of these emanations without being affected by them, as if in the purest atmosphere'.[140] Occupational illnesses could thus be conceived and euphemised as merely transient pathologies, linked to the demands of acclimatisation.

Taking this climatic perspective, hygienists were interested in the 'bodily changes' produced by the artificial climate of the factory. For example, after a few years, the skin and bones of copper workers became 'greenish or bluish'.[141] One doctor noted that white lead workers' 'skin was dyed so red for it to be possible to recognise them a year later'.[142] Logically enough, the forensic doctors of the 1860s studied these bodily transformations (the metamorphoses of the hand in particular), to identify bodies altered by various occupations.

The sole purpose of the 1810 decree had been to manage the conflict between industrialists and the owners of real estate; the problem of workers' health was carefully sidelined. Or rather, the liberal logic that governed financial compensation for environmental damage also applied to workers' bodies. Jurisprudence in the first part of the nineteenth century relied on the old Smithian theory of compensatory wages (higher wages made up for the greater risks) to refuse to pay compensation to

---

139   François Mêlier, 'De la santé des ouvriers employés dans les manufactures de tabac', *AHPML*, 34, 1845, pp. 241–300.

140   Alphonse Dupasquier, 'Mémoire relatif aux effets des émanations phosphorées sur les ouvriers employés dans les fabriques de phosphore', *AHPML*, 36, 1846, pp. 342–56.

141   Ambroise Tardieu, 'Mémoire sur les modifications physiques et chimiques que détermine dans certaines parties du corps l'exercice des diverses professions pour servir à la recherche médico-légale de l'identité', *AHPML*, 42, 1849, pp. 388–423.

142   Brechot, 'Mémoire sur les accidents', p. 77.

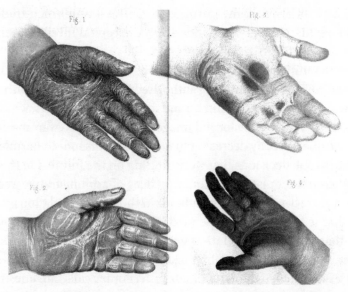

Maxime Vernois, *De la main des ouvriers et des artisans* (Workers' and Tradesmen's Hands)[143]

sick or injured workers. Thus, according to a ruling by the Lyon court of appeal in 1836, workers were not entitled to claim compensation because 'the risks that their work may present are compensated . . . by the special wage for their type of occupation'.[144] Through the employment contract, the worker had supposedly recognised and accepted the risks of the job. And in the techno-liberal logic of the time, it was assumed that such compensation for risks through wages would encourage the entrepreneur to improve safety: according to the engineer Freycinet, safer conditions paid for themselves, since workers who took greater risks demanded higher wages.[145] Until the end of the nineteenth century, there was no law to stand in the way of this logic.

We cannot think about the modern destruction of the environment without thinking about the history of power. The chemical industrialisation that took place between 1800 and 1830 was made possible by the political transformations that followed the French Revolution. These extraordinarily polluting factories and the new environmental regulations they necessitated were imposed by the government, in the name of national prosperity, against the interests of residents and for the benefit of a small clique of manufacturers close to those who held power. It was

143   *AHPML*, 2nd series, 17, 1862.
144   *Journal du Palais*, Paris, 1836, ruling of 29 December 1836.
145   Charles Freycinet, *Traité d'assainissement industriel*, Paris: Dunod, 1870, p. 6.

from the exception that the norm was born, at a time when traditional forms of regulation were suspended and the capital invested seemed to make it impossible to 'turn back the clock'.

After the initial coup of the 1810 decree, two different logics clashed: that of the administration, which authorised factories in accordance with a nationwide industrial programme; and that of the civil courts, which arbitrated on the value of the local damages which this programme caused.

On the one hand, by decreeing that a factory was indeed harmless, the authorities took decisions that were binding on the future. On the other, the civil courts regulated the consequences of administrative decisions *a posteriori*. Technology was caught up in these power relations; and the forms that it took closely depended on society's ability to mobilise to ensure that the price of pollution was paid.

Of course, the fact that a financial form of pollution regulation was already established in the early nineteenth century calls into question the pertinence of the current dominant approach to environmental problems. Neoclassical economics formalised the idea that nature has a price, or that a price must be set for it in order to reach an economically optimal point of pollution – that is, a fair allocation of resources, answering both the search for economic efficiency and the demands of environmental protection. Yet this idea corresponds to an old, indeed long-generalised, practice of compensation for environmental damage.

It is clear that this mode of environmental regulation has not prevented pollution; on the contrary, it has historically accompanied and justified the degradation of the environment. In fact, there is an intrinsic logic to this regulation, the consequences of which have been apparent since the 1820s. The principle of compensating for damage, combined with the imperative of economic profit, produced three results: first, the hiring, for the most dangerous tasks, of the most vulnerable populations, whose hardships could remain socially invisible; second, the concentration of production and pollution in a few localities; and third, the choice to situate them, in particular, in poor territories which lacked the social and political resources that would increase the value of environmental compensation. We can only conclude that this logic still holds today, and that it has undoubtedly become even more pronounced as a result of economic globalisation.

5

Lighting Up France after Waterloo

The idea of a technical safety standard purports to make the world safe by defining technical forms that must be followed. Yet such a standard presupposes the ability to discipline and control the varied world of objects. If, today, such a practice is commonplace, it also needed inventing. More specifically, it was invented at the intersection of the French administrative and academic worlds in the 1820s, in response to the breakthroughs which British technologies had made during the Industrial Revolution. In 1823, the French government determined that steam engines and gasholders (which had not yet caused any accidents in France) would have to obey a special form defined by the Académie des sciences. This was a new and radical political act: the government believed that science could make the world of production safe by defining rational technical forms in advance.

In the eighteenth century, safety standards, as stipulated by the police or guilds, had been based on the experience of communities of tradesmen. They applied at a local level, since they were linked to the institutions and customs of particular towns; and they were developed through case law, in response to accidents or reprimands for shoddy workmanship. Specifying what should *not* be done, regulations were based on the observation of malpractice rather than on any theory of how things ought to be.[1] The 1820s brought a considerable change in this regard: risk, which had previously been managed by the urban practices of the police, now belonged to the realm of the scientific institutions. Where the police offered a day-to-day regulation of safety issues, the post-revolutionary administration sought to establish advance guarantees that new and little-known technologies would behave in a perfectly reliable fashion.

---

1   Robert Carvais, *La Chambre royale des bâtiments. Juridiction professionelle et droit de la construction à Paris sous l'Ancien Régime*, PhD in Law, Paris II, 2001.

Scientific standardisation was possible because it emerged in the vacuum created by the suppression of the guilds and the rise of new technologies that lay beyond the know-how of established trades. This discontinuity allowed the government to impose practices inherited from other worlds: the control of products by administrators (such as the French factory inspectors or the British excise's quality specifications), or control by engineers who sought to define the best possible technical solution mathematically.[2] The idea that there did indeed exist a technical optimum was crucial, for it provided the underpinnings of governments' pretention to impose a single standard throughout the country.

Standardisation in the 1820s also fitted closely with the notion of *technology*. This idea, which was in fact new at the time, held that it was possible (through drawing and text) to set technical devices down on paper, and provide a representation of them sufficiently exhaustive to ensure their proper functioning. In this respect, this notion bears some similarities to patents. Under the Ancien Régime, the so-called 'privilege of invention' required that the object be *presented* (before the prince or a learned institution) and it had to be demonstrated in operation. Yet the 1791 law on patents required only a *representation* of the invention on paper. Standards and patents were part of the same political project, which sought to extract 'the idea' from the technique and circulate it for the greater public good.[3] With scientific safety standards, the French administration also sought to improve the state of industry throughout the nation.

Such standards would ultimately emerge in response to a major conflict which pitched industrial capitalism against those who drew rents from property. In Paris, gas lighting was one of the main battlefields. Since gasometers endangered certain upmarket districts with the threat of explosions, the residents of these areas fought the industrialists hard.

---

2   Philippe Minard, *La Fortune du colbertisme. État et industrie dans la France des Lumières*, Paris: Fayard, 1998; William J. Ashworth, 'Quality and roots of manufacturing expertise in eighteenth-century Britain', *Osiris*, vol. 25, no. 1, 2010, pp. 231–54. Ken Alder, 'Making things the same: representation, tolerance and the end of the Ancient Regime in France', *Social Studies of Science*, vol. 28, no. 4, 1998, pp. 499–545; Eda Kranakis, *Constructing a Bridge: An Exploration of Engineering Culture in Nineteenth-Century France and America*, Cambridge, MA: MIT Press, 1997. For a different view of what 'optimisation' means, see Frédéric Graber, *Paris a besoin d'eau*, Paris: CNRS éditions, 2010, p. 362: 'The best, as practised by engineers, consists in considering only one project: the final project will be the best, because it will be *the only one*.'

3   Jacques Guillerme and Jan Sebestik, 'Les commencements de la technologie', *Thalès*, 12, 1966, pp. 1–72; Liliane Hilaire Perez, *L'Invention au siècle des Lumières*, Paris: Albin Michel, 2000; Mario Biagioli, 'Patents republic: representing inventions, constructing rights and authors', *Social Research*, vol. 73, no. 4, 2006, pp. 1129–72.

The aim of the first safety standards was to reach a compromise which would rely on technical fixes. Thanks to these standards, risk was both controlled and legalised.

Lastly, standards were a typically French way of colonising the future. Britain, although familiar with the dangers of gas and boiler explosions, did not opt for standardisation. In 1817, the British government rejected a bill which sought to standardise steamboat boilers. Despite repeated disasters, not until 1852 would maritime boilers be regulated. Neither boilers on land nor gasometers were standardised in the nineteenth century. In France, by contrast, maritime and land boilers, as well as gasometers, were subjected to precise safety standards from 1823 onwards. British industrialists, engineers and MPs considered such a solution rather presumptuous. To their eyes, the safety of technical devices ultimately depended on innovation, and it was better to give free rein to the inventiveness of profit-driven engineers, rather than rely on standards that would become obsolete as soon as they were issued.

1. Technology in the public space

Gas lighting was one of the most controversial innovations of the Industrial Revolution. There was sure to be controversy in replacing the fireplace – symbol of the private sphere – with gas spouts linked to industry by a vast technical grid, and indeed in building huge gasometers in Paris which contained millions of litres of flammable gas. Yet the story of this dispute – which raised questions about the choice of coal, the engineering infrastructure required, and the concentration of risks – has near-exclusively been told from the perspective of industrial history. In other words, it has been told with scant regard for the debates that took place, or for those who opposed this development. Successive accounts of the controversy have, instead, built up an image of reactionary and irrational opponents. As early as 1843, Adolphe Trébuchet, a member of the Paris Conseil de salubrité, explained that this institution had successfully defended gas, despite the panic among Parisians.[4] In 1872, Louis Figuier held forth at length on the 'strange aberrations' of opponents of gas, said to 'offend reason'[5]; and even a century later, the economic history was hardly more

---

4   Adolphe Trébuchet, *Recherches sur l'éclairage public de Paris*, Paris: Baillière, 1843, p. 49.

5   Louis Figuier, *Les Merveilles de la science*, Paris: Jouvet et Cie, 1872, vol. 4, p. 130.

charitable.[6] Oddly enough, all these accounts gloss over what was actually being debated: critics were mainly opposed to the installation of gasometers in the middle of residential areas, insisting – despite industrialists' and experts' denials – that they risked explosions. Since it did not take long before gasometers did indeed explode, it has to be said that the opponents proved to be correct.

The gas lighting controversy is above all remarkable because it gripped the public arena. As far back as smallpox vaccination, no innovation had aroused such passions in France: newspaper articles, pamphlets, pop-science, advertisements, a play and even an opera were written either in defence or criticism of lighting by gas.[7] Unlike steam engines or chemical factories, which were relegated to outlying industrial districts, gas lighting conquered fashionable settings from theatres to opera houses, restaurants, cafés and reading rooms. Frederick Winsor, alias Winzer, a German merchant who had set up the first lighting company in London in 1809 (and then gone bankrupt) decided to publicise gas by lighting the Passage des Panoramas free of charge. Located between the Palais-Royal and the Grands Boulevards, this covered arcade symbolised a commercial, comfortable and elegant urban modernity. Winsor fitted out a small showroom where the public could obtain the latest brochures and compare the lighting power of gas with that of oil lamps.[8]

In the early 1820s, innovation was still part of the culture of the public sphere. The industrial exhibitions that had been held regularly in Paris since 1798 were modelled on the great paintings exhibitions organised each year in Paris: they were intended to give the public an opportunity to exercise its critical faculties.[9] One of the organisers of the 1819 exhibition explained that the aim was to bring the latest inventions together in one place so that the public could compare and judge them.[10] The many pamphlets on gas followed in this tradition of valuing the public's judgement. Nicolas Clément-Desormes was one of the few chemists to

---

6   Jean-Pierre Williot, *Naissance d'un service public, le gaz à Paris*, Paris: Rive Droite, 1999.

7   Ferdinand Léon, *Le Magasin de lumière, scènes à propos de l'éclairage par le gaz*, Paris: Mme Huet, 1823; *Aladin ou la lampe merveilleuse*, Paris: Roullet, 1822.

8   Frederick Winsor, *Résumé historique et démonstratif sur l'éclairage par le gaz hydrogène*, Paris: Didot, 1824, p. 10.

9   Translator's note: The Salon was the exhibition long held by the Académie royale de peinture et de sculpture (later the Académie des beaux-arts). From the late seventeenth century it was the central event in Paris's art calendar, showing the works of Académie graduates; from 1737 it was open to the general public.

10   Victor de Moléon, 'Discours préliminaire', *Annales de l'industrie nationale et étrangère*, Paris: Bachelier, 1820, p. 43.

speak out against gas. In his view, since the consequences of innovations affected society as a whole, the assessment of their usefulness had to be made by as wide and varied a public as possible: 'Unless there is a public debate that draws the attention and contribution of a great number of persons, no sound judgement can be made.'[11]

2. Controversy as a means of assessing technology

During the controversy, five aspects of the gas were debated: its economic consequences, the ugliness of its light, its insalubrity, the dependence of the individual and, finally, the danger of explosions.

At first, the debate was purely economic. Like any innovation, gas seemed to threaten whole sectors of the economy. In London, its opponents argued that it would lead to the bankruptcy of lighting oil merchants and, in turn, whalers. Given that whalers were reputed to be the kingdom's finest sailors, this innovation also appeared to threaten British maritime dominance and national security.[12]

A pamphlet published in Paris in 1816 castigated gas for being an 'anti-national' technology.[13] The author predicted the ruin of rapeseed growers, oil purifiers and lamp manufacturers – a ruin that was likely to heighten popular discontent and lead France to a new revolution. On a more serious note, Nicolas Clément-Desormes examined the economic context of the innovation: while it was profitable in London and Manchester, it was ill-suited to the French economy, where both coal and its transportation were three times more expensive.[14]

Lighting gas raised the question of the desirable model of economic development should be. Could – should – France follow the same path as Britain? According to the *académiciens*, the main advantage of gas was that it would encourage the growth of the coal industry: demand would create supply, coal prices would fall, canals would be dug, and France would follow Britain down the path of coal-based industrialisation.[15] This debate went hand in hand with discussions of the stewardship of natural

11   Nicolas Clément-Desormes, *Appréciation du procédé d'éclairage par le gaz hydrogène du charbon de terre*, Paris: Delaunay, 1819, p. 2.

12   Frederick Winsor, *Plain Questions and Answers Refuting Every Possible Objection against the Beneficial Introduction of Coke and Gas Lights*, London: G. Sidney, 1807.

13   *Réclamation de l'huile à brûler contre l'application du gaz hydrogène à l'éclairage des rues et des maisons*, Paris: Locard, 1817.

14   Clément-Desormes, *Appréciation du procédé d'éclairage*, p. 8.

15   *Procès-verbaux des séances de l'Académie (des sciences)*, vol. 8, 2 February 1824.

resources: was it really reasonable to burn coal for the sake of lighting? Given that coal was used to produce steel – a product far more useful for national defence – did not gas lighting amount to sacrificing future generations' security for the sake of comfort today? Clément-Desormes distinguished between renewable and non-renewable resources: in order to develop French industry in the long run, it was necessary to save coal, which, unlike rapeseed oil, 'cannot be reproduced indefinitely'.[16] Chaptal concurred that it was better to use the national coal stock to produce steel.[17] The advocates of gas also considered the long term: the coke produced by distilling coal would be used for domestic heating and would help to conserve forests. Coincidentally, it was during the meeting of the Académie des sciences following the discussion on gas lighting that its members studied the link between deforestation and climate change.[18]

In 1823, the Romantic writer Charles Nodier and the doctor Amédée Pichot published their *Critical Essay Against Hydrogen Gas*. The innovation was accused of making the world uglier by removing shadows and their mystery, and of infecting Paris by introducing swamp air. Was not the gas that came out of the nozzles similar to that which produced will-o'-the-wisps and epidemic fevers? Distilling coal, that is, organic residues which had accumulated over centuries, would reintroduce age-old putridity, buried under protective soil, into the heart of the capital.[19] Since the establishment of gasworks, the withering of trees along gas pipelines, fish found dead in the Seine and the disappearance of swallows in London were so much evidence that the danger was real.[20] Gas also contains sulphides that attack the lungs. Dr Pichot studied the process in some detail to show that gas could not be perfectly purified: if the gas remained in contact with the washing water for too long, it would end up purified but no longer produce light. Distillation was just as tricky: coals of varying qualities, with varying levels of sulphur, made the presence of sulphurous gas inevitable.

A danger of a wholly different order came from the loss of control over lighting, which now depended on a vast network of engineering infrastructure. There was major reticence over such dependence.[21] Conversely,

16   Clément-Desormes, *Appréciation du procédé d'éclairage*, pp. 38–40.

17   Jean-Antoine Chaptal, *Quelques réflexions sur l'industrie en général, à l'occasion de l'exposition des produits de l'industrie française en 1819*, Paris: Corréard, 1819, p. 16.

18   *Procès-verbaux des séances de l'Académie des sciences*, vol. 8, 2 February 1824.

19   Charles Nodier and Amédée Pichot, *Essai critique sur le gaz hydrogène*, Paris: Gosselin, 1823, p. 31.

20   APP, RCS, 26 May 1823; *Des Dangers de l'existence des gazomètres en ville*, 1823, p. 3–6; *Gentleman's Magazine*, 93, 1823, p. 224.

21   *Réclamation de l'huile à brûler*.

'Do not confuse me for anything that has appeared before'
(Musée Carnavalet, prints, © PMVP/Cliché: Joffre)

one 1823 advertisement tried to show that gas would keep a man in control of his home: thanks to gas, in a moment, the courageous husband could jump out of bed, light his lamp and point a gun to the robbers, thus protecting the marital bed.

In public settings, sudden darkness was even more threatening. Nodier and Pichot imagined the panic in a theatre suddenly devoid of lighting:

> All of a sudden, the lighting equipment . . . goes out and leaves the audience in deep darkness. With one hand on my watch and the other on my purse, I escaped amid cries of terror, admiring the ingenious instinct of the police, who had entrusted all the fortunes of public safety to the caprice of some simultaneous light.[22]

The murder of the Duc de Berry (whose potential offspring was the last hope of continuing the Bourbon lineage) by the Bonapartist Louvel as he left the Opéra in February 1820 was still fresh in everyone's memory. Opponents of gas insinuated that this innovation could be the starting point of a plot by opponents of the monarchy: 'Suppose all the royal family at the Opéra, admit the inadvertence of a gasometer worker, and suppose

---

22   Nodier and Pichot, *Essai critique sur le gaz hydrogène*, pp. ix–x.

a Louvel in the crowd amidst the darkness !!!! . . . And what means a *party* would have to obtain this inadvertence.'[23]

The use of gas for insurrectionary purposes was taken very seriously: in London, a House of Commons select committee debated at length whether insurgents could explode a gasometer.[24] In 1820s Paris, the scene of repeated attacks and plots against the king (no fewer than ten such attempts were foiled between 1820 and 1823), the new method of lighting, with its underground pipes carrying an invisible explosive substance, seemed to make police surveillance techniques obsolete.[25] Even if these fears were exploited to damage the liberal party – already accused of permissiveness after the assassination of the Duc de Berry – the fact remained that gas lighting seemed to pose a considerable political and technical risk in a Paris that had not yet been well pacified.

3. An honourable monument to French industry

Up to 1823, the issue of explosions had remained confined to administrative circles: the Conseil de salubrité and the offices of the Interior Ministry. But suddenly the controversy boiled over. In question was the existence of an enormous 200,000 cubic-foot gasometer that Antoine Pauwels and his Compagnie française du gaz d'éclairage had installed on the rue du Faubourg-Poissonnière, at the time a fashionable district on the Right Bank, close to the boulevards and arcades.[26]

In 1823, this gasometer was an utterly extraordinary technical and political object. Technically, it was quite monstrous. In Paris, gas lighting was still in its infancy: the small plant set up by Winsor to light the Senate had faced setbacks, and the apparatus installed by the académiciens Gay-Lussac, Girard and Darcet at the Hôpital Saint Louis was no more than a prototype. Pauwels's gasometer was ten times larger than the largest ones in Britain, built by engineers with far more experience. During the House of Commons' hearings on gas safety, an engineer termed this apparatus almost impossible to control.[27]

---

23   *De l'Éclairage par le gaz hydrogène*, Paris: Dondey-Dupré, 1823.

24   'Report of the select committee on gas-light establishments', *House of Commons Papers, Reports of the Committees*, 1823, vol. 5, pp. 7, 13, 32.

25   Berthier de Sauvigny, *La Restauration* (1955), Paris: Flammarion, 1999, pp. 180–3.

26   Pascal Etienne, *Le Faubourg Poissonnière*, Paris: DAAVP, 1986.

27   'Report of the select committee on gas-light establishments', p. 26.

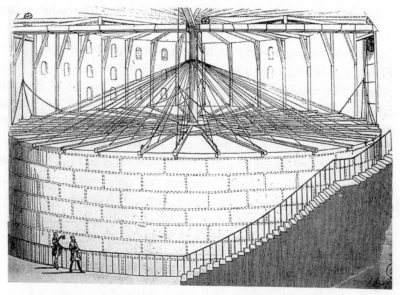

The gasometer of the Compagnie française du gaz d'éclairage

The gasometer was also explosive in political terms. In 1821, the liberal government of the Duc Decazes ceded power to the ultra-royalists. However, the gasometer authorised by the previous government had been financed by members of the liberal party, nobles of the Empire and Freemasons – in short, everyone that the ultra-royalists loathed. In 1823, Pauwels was a twenty-seven-year-old self-taught engineer who had served as a pharmacist in the imperial armies. After the Restoration, he manufactured various chemical products in Paris before raising 3 million francs to build the largest gasometer in the world. His shareholders included Laffitte, a major banker close to the liberal party; the Duc d'Orléans, a dynastic rival to the ultra-royalist Duc d'Artois; the Duc Decazes, head of the fallen government; d'Anglès, a former protégé of Fouché, who had become prefect of the Paris police and who had authorised the gasometer's construction; Boulay de la Meurthe, a minister during the Hundred Days and a famous outlaw; Saint-Aulaire, a deputy from the liberal party; and Manuel, a notorious freemason and member of the Charbonnerie secret society.[28] So when, in September 1823, the Conseil d'État overturned the administrative authorisation granted to Pauwels by d'Anglès, the liberal press cried scandal: the government had merely used the pretext of security to act according to partisan motives. The largest gasometer in the world, this 'honourable monument

---

28   List of shareholders in *Des Dangers de l'existence des gazomètres en ville.*

to national industry' was threatened with destruction for sordid political reasons.[29]

The gas explosions reported in the press (seven in London and two in Paris between 1819 and 1823) seemed to indicate that such mishaps were not, indeed, impossible. The opposition to Pauwels was led by the influential Baron Charles Athanase de Walckenaer, an alumnus of the École polytechnique, master of requests at the Conseil d'État and member of the Institut. In his petitions, he developed a subtle argument reminiscent of our precautionary principle: a gasometer explosion was surely unlikely, but the consequences would be so severe that no uncertainty could be tolerated. It was thus necessary to ban gasometers in Paris.

The stakes were clearly raised by Nodier and Pichot, who imagined Paris razed to the ground: Pauwels was a 'manipulator who held the lives of six hundred thousand citizens at the mercy of an error'. They were outraged that the administration 'had dared to put the lives of a million men at the mercy of a demented act or a surge of despair'.[30] An initial report by the académiciens had recognised that 'an explosion would be a disastrous event for the whole neighbourhood', but considered this 'absolutely improbable' and even 'a chimera'.[31] Conversely, opponents insisted that the likelihood of an accident remained unknown and that, as a result, insurance companies were refusing to insure the gasometer or their homes. They challenged the members of the Conseil d'État to take moral responsibility for the inevitable scientific doubt: 'The académiciens have told you: "The explosion is possible but it is unlikely if we take the means that prudence and science suggest." We would ask you, Gentlemen, if these may not prove lacking, and if you will take such responsibility on your consciences.'[32]

On 10 September 1823, the Conseil d'État overturned the authorisation that d'Anglès had granted to Pauwels in 1821, recognising that this was too important a decision to be taken through a simple prefectoral decree. Responsibility was transferred to the government, which had to decide in which class of the 1810 decree gasometers should be ranked. If it was to be the first class, Pauwels would simply have to dismantle his plant and rebuild it outside Paris, far from any dwellings. To resolve the

---

29   Antoine Pauwels, *Notice sur le grand gazomètre de la rue du Faubourg-Poissonnière*, Paris, 1823, p. 33; *Journal du commerce*, 23 September 1823. For this article, the newspaper's editor was sentenced to three months in prison.

30   Nodier and Pichot, *Essai critique sur le gaz hydrogène*, pp. 4, 97.

31   AD Seine, V8 01 1002, Rapport de Cordier, Thénard et Gay-Lussac sur le gazomètre de Pauwels.

32   *Des Dangers de l'existence des gazomètres en ville*, p. 7.

matter, the government commissioned the Académie des sciences to draw up a 'solemn report'.

4. French *académiciens* and British expert witnesses

Lighting gas provides an ideal case study for comparing expert appraisal practices in France and Britain. In the two cases, highly contrasting forms of expert appraisal were being carried out at the same time, on the same subject: academic reports on the one side of the Channel, and parliamentary hearings on the other.

At the end of 1823, the aim of French Academy of sciences report on gas was twofold: to put an end to a controversy that was becoming polit- ically embarrassing, and to save the Pauwels gasometer. Castelblajac, the director of commerce and manufactures, gave the Académie des sciences detailed instructions. The destruction of the gasometer was never consid- ered a serious possibility. He asked only for safety measures – 'you will consider gas lighting, its dangers and their remedies' – as if one could not exist without the other.[33]

The *académiciens* Gay-Lussac, Girard, Darcet, Héron de Villefosse and Thénard, who were appointed to write the report, took an analytical approach. In a series of paragraphs, they described the risks involved in coal storage, gas production, gas purification, gas storage and the distribution of gas through pipelines, without considering the cross- effects between the various factors (such as the distillation of coal next to a gasometer). Gasometers were reduced to a simple physical model, whose safety could be inferred from the certainty of the laws of physics. For example, gas explodes only when properly mixed with four to twelve times the same volume of air. But in gasometers, the gas is always under pressure, and thus air cannot make its way into the vessel. Gas is also much lighter than air, so it would take several hours for them to mix. Nor was it necessary to fear lightning: given that gas is not an electrical conductor, the electrical discharge would be conducted through the bell of the gasom- eter into the water-filled tank. Finally, gas pipes would not communicate flames and carry the fire over any distance; for scientists had shown that a flame cannot pass through a pipe longer than its own diameter. The *académiciens* then reasoned according to the theory of probabilities. The 'gasometer explosion' event was conceived as a combination of sub-events,

---

33   AAS, pochette de séance, 9 February 1824. Letter from Castelblajac to Corbières, sent in CC to Cuvier.

namely, a depressurisation of the gasometer, an opening in the vessel, a long period without supervision for the mixture to mix properly and, finally, a flame being present in the mixture. Since each of these sub-events was considered highly improbable, a combination of all of them was deemed 'absolutely improbable' or 'a chimera'.[34]

The British experts did not follow a similar logic. For example, William Congreve, a leading explosives expert, did not offer thought experiments, but instead investigated the accidents that had already taken place, which proved highly instructive. For instance, an explosion at the Great Peter Street gasworks in Westminster had been caused by the heat of a distillation retort placed too close to the gasometer.[35] The status of experiments was also very different in each case. The British experts questioned by the select committee often refused to answer, for want of experimental evidence. The laboratory experiment seemed too far removed from the real situation at hand. According to the chemist Humphrey Davy: 'From experiments made on a small scale, it is not possible to reason with perfect confidence when the scale is 100,000 times larger.'[36] This difference in scale between the laboratory and industry was extremely worrying because it was feared that the gas's explosive power might increase exponentially. Experimental practices were themselves rather different. The British were more pragmatic: when the committee asked whether the explosion of a gasometer could cause the explosion of other gasometers placed near it, Davy suggested building a scale model of a factory and exploding a miniature gasometer. Congreve, to determine the effect of an explosion of gas, built a gas cannon: from its range he could infer an equivalence between the explosive powers of gas and gunpowder.

But when the same problem was investigated in the Académie des sciences, it became a classical question of experimental physics. Ampère proposed studying the explosive power of gas because 'we would at the same time solve a question of physics of great interest: the determination of the heat produced in the combustion of various inflammable gases'. He then explained how the danger of explosions could be entirely deduced from the temperature of the burning gases. Pelletan, who responded to Ampère's proposal, sent a letter to the Académie setting out his initial results. The initial problem – gasometer explosions – was no longer even mentioned. The expert's approach had been transformed into a

---

34   *Procès-verbaux des séances de l'Académie (des sciences)*, vol. 8, 2 February 1824.
35   'Report of the select committee on gas-light establishments', p. 61.
36   Ibid., p. 7.

pure problem of physics, using a classic experimental protocol and not producing any concrete results.[37]

The status of subjectivity, doubts, and contradictory evidence varied greatly between British and French expertise because of the distinct judicial traditions of each country: an oral and adversarial procedure in England, and a written and inquisitorial one in France. In France, the government commissioned a written report, which could remain confidential. In Britain, the expert was heard as a witness, his evidence was oral and subject to contradiction.

The British parliamentary select committee on gas organised hearings with a wide range of figures: scientists, gas engineers, London company directors and even some opponents of the plants. Often, the supposed objectivity of the expert gave way to the expression of his personal opinion. For example, the chemist Humphry Davy was not questioned about his expertise on gasometers: he admitted that he had never seen one before the morning of his hearing![38] What the select committee expected from Davy was his subjective and personal assessment of the risk:

– There are fourteen gasometers placed very near together; what do you conceive the effect of the explosion of one of the largest gasometer in Peter-street would be, amongst the other gasometers?
– It is extremely probable that they would be overturned, torn in pieces and the whole exploded but it is a matter of experimental research . . . nobody can predict the exact result.

When the committee insisted on having an answer, he gave a personal opinion:

– Upon the whole, would you not rather they should be farther apart?
– Certainly, if I were living in the neighbourhood I would certainly rather live in a neighbourhood where they were further apart.
– Would you feel uneasy from their situation, if you lived in the neighbourhood?
– I think I should.
– You have stated that if you were to exercise your own choice, you would not place your residence in the immediate neighbourhood of the gas works?
– Unquestionably not.

---

37   AAS, pochette de séance, 1 and 8 March 1824.
38   'Report of the select committee on gas-light establishments', p. 9.

Samuel Clegg, a prominent expert in gas lighting, tried to counter the bad impression made by this statement by asserting that he 'should have no objection to [having his own] bed placed on the top of a gasometer; [he] should sleep there as in any other place'. As for a gasholder explosion, he 'thought it quite impossible and should as soon guard against the falling of Waterloo Bridge from an earthquake, as guard against any explosion of that sort'.

When, as in Clegg's case, an expert seemed very sure of himself, cross-examination could introduce doubts. A neighbour of the St Pancras works managed to push Clegg onto the retreat:

– You have been asked by the committee whether the gasometers at St Pancras are safe?
– Yes.
– Have you viewed them?
– They are not erected yet, but I have seen the plan.
– Do you know the size of those gasometers to be erected?
– No.
– Do you know the distance of the gasometers from each other?
– Not exactly.
– Do you know how many there are?
– I do not know, three at present I believe.
– Do you know the distance the gasometers are now to be placed from the houses?
– No I do not.

The procedure followed by the French *académiciens* managed risk in a totally different way. First, unlike the oral, contradictory hearing, their written report enabled them to present a single opinion that seemed perfectly coherent. Yet the draft of the report, held in the archives of the Académie des sciences, shows that disagreements were indeed present throughout the process of producing the document. It is an outright battlefield, written by several hands and in several colours. The pasted-together sheets show the evolution between different versions. Drafting techniques (erasing, pasting, crossing out, adding, snipping) obliterate the experts' train of thought, which is full of detours, objections and uncertainties. Conversely, in London, the *Minutes of Evidence*, which published the debates in detail, lifted the veil on this preparatory phase.

The draft report was part of a wider set of techniques in the Académie designed to produce a consensus. On many points, two *académiciens*,

Thénard and Héron de Villefosse did express disagreement. But these dissenting voices had a hard time breaking through the fundamental barrier between the spoken and the written word: Thénard ultimately fell in behind the majority opinion and settled for reading out a note on additional safety measures. Héron de Villefosse, who refused to sign the report, was a man whose opinion the Académie could not easily ignore: close to Louis XVIII, who made him a baron in 1820, he was secretary to the king's cabinet and a Conseiller d'État. However, the minutes only mention his opinion in a brief euphemism: 'M. Héron de Villefosse, a member of the commission differing *in some respects* from the opinion of his colleagues, read an *individual* report on this subject.' His paper was not put to the vote of the *académiciens*, who could only pass judgement on the majority report. Nor was the debate that took place on 9 February transcribed: a brief note listing the speeches indicates that Walckenaer, who was present during the debate, made 'a few miscellaneous observations'.[39] Finally, the tally of votes for and against the majority report was not published. In short, the Académie's procedure was perfectly built for masking disagreements and manufacturing unanimity and present a solid foundation for the government's decision to authorise Pauwels's enormous gasholder.

Yet the voice of science, as constructed by the Académie system was far from neutral, disinterested or objective. Seen in retrospect, the conclusion of the report was easy to foresee, for many *académiciens* were interested in gas for financial, intellectual, and even patriotic reasons. In 1823, Gay-Lussac was a consulting chemist for Manby and Wilson, who had just set up a gasworks at the Barrière de Courcelles. As early as 1817, he had travelled to London, accompanied by other *académiciens*, to study the new method of lighting, in the hope of establishing a company. He even offered to light the Senate, but the project was instead entrusted to Winsor.[40] Along with Darcet and Girard (who were also rapporteurs), he took part in the construction of the lighting apparatus for the Hôpital Saint-Louis. Girard became chief engineer of the Compagnie royale d'éclairage par le gaz, founded by Louis XVIII to light the Opéra. In 1822, the company was taken over by Chaptal fils in partnership with Darcet.[41]

---

39   AAS, pochette de séance, 9 February 1824.

40   APP, RCS, 17 November 1817.

41   AN F12 2269, AN O3 1588; AN F12 196 bis, 12 February 1824, 'Avis du Conseil général des manufactures sur le gaz'; AAS, dossier Girard. We have seen the business relations between Chaptal fils and Darcet in the case of the Ternes factory in Neuilly and the Plan d'Aren factory in Istres.

Gas lighting was also of great scientific interest: Girard and Navier made experiments and constructed intricate differential equations to describe the flow of gas through long pipes. This scientific involvement was typical of post-Waterloo French science: the *académiciens* thought that their social role was to help France catch up with British technological advance. They considered that every technology of that period coming from England could be improved and 'perfected' with applied mathematics.[42] This conception of a potentially perfect technology, obtained through scientific expertise, also drove the *académiciens'* technological optimism about the risks of gas lighting.

Finally, it must be emphasised that these contrasting practices of expertise produced uneven results. By interviewing the engineers who worked on the gasometers on a day-to-day basis, the Commons select committee gathered concrete information on practical difficulties. Engineers were concerned about the heat in the distillation retorts, which caused the gas in the gasometer to expand or even burn; the limewater used to purify the gas could evaporate, letting the gas escape; and the wind could cause the gasometer vessels to sway or even topple over.

When the *académiciens* visited the Pauwels plant, they did not notice any of these problems. Fortunately, before drafting their report on the classification of gasometers, they were able to consult the hearings of the British select committee and plagiarise its proposals. The royal decree of 28 August 1824 endorsed their report: gasworks were placed in second class and there was no limit to the size of gasometers. In return, entrepreneurs had to comply with the safety measures prescribed by the *académiciens*.[43]

5. Norms and the legalisation of risk

The 1824 decree enshrined Pauwels's victory, and his huge gasholders remained authorised. Better still, he had succeeded in setting a precedent: the law for the whole of France was designed to suit him. Since no distance was required and the size of the gasometers was left open, the decree authorised the construction of gigantic gasholders in the middle

---

42   Pierre Simon Girard, *L'Écoulement uniforme de l'air atmosphérique et du gaz hydrogène carboné dans des tuyaux de conduite*, Paris: Feugueray, 1819; Claude Louis-Marie Navier, 'Sur l'écoulement des fluides élastiques dans les vases et les tuyaux de conduite', *Annales des mines*, 6, 1829, pp. 371–442.

43   Jean-Baptiste Fressoz, 'The gas-lighting controversy: Technological risk, expertise and regulation in nineteenth-century Paris and London', *Journal of Urban History*, vol. 33, no. 5, 2007, pp. 729–55.

of every town and city in France. The safety standard was defined in obedience to a technological object deemed monstruous by many experts.
So, how are we to explain the paradox?

More than all the academic reasoning, it was the network of technical
infrastructure that ensured its own victory. The construction of huge
metal structures and the burial of pipes in the ground rendered any
subsequent changes extremely costly. For example, simply moving these
structures outside the towns would mean changing the entire network,
because the diameter of the pipes gradually decreased from the gasometer
to residents' homes.[44] Following the Conseil d'État's decision to withdraw
administrative authorisation, the *Journal du commerce* had challenged
the authorities to reimburse not only the industrialists involved, but
also shopkeepers and restaurant owners who had made investments in
anticipation of the arrival of gas.

Too many people had invested in gas: the consumers whose properties
had already been fitted out, the manufacturers of the equipment used, the
financiers, the administrators and the experts who had supported this
technique. The Académie des sciences, for example, was too committed to
back down. An anonymous note recalls that, at the end of the eighteenth
century, it had encouraged the inventor of gas lighting, Philippe Lebon,
in his project, and that it could hardly go back on its word, especially now
that the British had turned this French invention into a lucrative industry.
The king himself owned shares in the Opéra lighting company.[45] Pauwels's
fate was thus bound up with past decisions by the administration and
the Académie, with the king, with competition with Britain, and with the
nation and progress itself.

There was also a perverse interaction between all the capital invested
and the principle of the universality of the law. In 1823, at the Académie,
the Pauwels factory was on everyone's mind. According to Héron and
Thénard, the Pauwels gasometer should be taken as a case apart, and
future factories banned from the towns and cities of France. But the
académiciens rejected this compromise, precisely in the name of the
general applicability of the law:

> The companies have committed six million on the faith of the govern-
> ment. It is obvious that the government cannot give retroactive effect to
> its acts. It would then allow existing establishments to survive, but there

---

44 'Report of the select committee on gas-light establishments', p. 49.
45 AAS, pochette de séance, 9 February 1823.

would henceforth be privileges for those establishments, and there would therefore be two different laws governing one same industry; that is where an ill-considered opinion would take us.[46]

The fact that the fate of Pauwels's gasholder was to be decided along with that of the general rules for gasworks was crucial: the spread of huge gasholders throughout the towns and cities of France was authorised for fear of damaging one jewel of national industry. Unlike the local police regulations of the Ancien Régime, the norms were supposed to be the same all across France. Enacted after the fact, this standard legitimised the industrial *fait accompli* in its most dangerous form.

6. The reign of the unpredictable

With the advantage of hindsight, how valid were the forecasts made by experts, proponents and opponents of the gasholders? In retrospect, what seems to prevail is uncertainty. The case of gas highlights the difficulty or even the impossibility of predicting the behaviour of complex technologies when they are plunged into the real world.

It was not long before gasholders did indeed explode, but none on the scale that their opponents had predicted. In 1844, the largest gasholder in Paris, containing 430,000 cubic feet of gas, suffered an accident described as 'very serious' by the authorities. Luckily, the huge flames fanned out into a field and 'only' one worker perished in the accident. The press reported the disaster with great concern. The prefect of police convened a commission of experts. Darcet, who had defended Pauwels in the early 1820s, made amends and recognised that the accident could have been much more serious.[47]

Another explosion in 1849, near the Opéra, finally convinced the most sceptical: even the *Journal de l'éclairage au gaz* stressed 'the importance for the city of getting rid of such an uncongenial and dangerous neighbour as the gasworks'.[48] Finally, in 1852, the minister of agriculture and commerce ordered that all new gasometers should be built outside towns.[49] But this tardy reaction did little to reduce the danger: given the urban sprawl of the 1860s, gasworks as far as La Villette in the northeast of Paris or Ivry in

---

46   Ibid.
47   APP, DA, 50.
48   *Le Journal de l'éclairage au gaz*, 20 April 1855.
49   APP, DA 50, circulaire du Ministre du commerce.

La Villette gas plant in 1878[50]

the south were closely surrounded by residential dwellings. The minister's decision thus amounted to protecting the capital and confining the risk to the industrial *faubourgs*.

In 1865, in London, the explosion of the gasometer at the Three Elms factory had more dramatic consequences: twelve dead, many injured and around a hundred houses devastated. The disaster stirred an emotional response among Londoners. The government had lied: 'They had been taught to believe that an explosion could not occur. They thought that they were as safe in the neighbourhood of gas factories as anywhere else; whereas they were living in constant peril.'[51] After the disaster, the press closed ranks in calling for gasometers to be banished from the city: 'At present it is clear every gasometer is a powder magazine'; 'The comfortable theory that a gasometer will not burst, must now be regarded as entirely exploded by the accident'; 'We should no more dream of accumulating explosive materials in quantities enough to blow an ironclad in the heart of [London] than of practising artillery in Oxford Street.'[52]

Opponents were also right when they denounced the toxicity of the gas. The danger was even greater than they had imagined: in 1823, the fear was that it was unhealthy and could cause asphyxiation. However, the first cases of gas poisoning showed that even small quantities could kill:

---

50   AD Seine, atlas 1007.

51   Thomas B. Simpson, *Gas-Works: The Evil Inseparable from Their Existence in Populous Places, and the Necessity for Removing Them from the Metropolis, as Has Been Done in Paris*, London: Freeman, 1866.

52   *The Times*, 6 November 1865; *London Review*, 4 November 1865; *Morning Herald*, 2 November 1865.

Explosion of the Three Elms Works gasholder[53]

gas was not only unbreathable – it was poisonous.[54] After the first cases of poisoning, forensic scientists studying the composition of lighting gas discovered a fearsome cocktail of ethylene, propylene and carbon monoxide. In the 1880s, Parisian gas contained 5–13 per cent carbon monoxide.[55]

As opponents had predicted, the purification of gas was always imperfect: at the beginning of the twentieth century, consumers were still complaining of bad smells, headaches and vomiting caused by gas. The large number of purification processes invented throughout the century, combined with the authorities' efforts to resolve what British newspapers called 'the sulphur question', clearly illustrate the difficulty of this task. In 1860, the London Gas Act set limits on sulphur levels, and numerous fines were imposed on companies that did not comply. The Board of Gas repeatedly had to raise these levels, as companies failed to respect the ruling.[56]

53   *Illustrated Times*, 11 November 1865.
54   Alphonse Devergie, 'Asphyxie par le gaz d'éclairage', *AHPML*, 3, 1830, pp. 457–75.
55   Alexandre Layet, 'Des accidents causés par la pénétration souterraine du gaz de l'éclairage dans les habitations', *Revue d'hygiène et de police sanitaire*, 2, 1880, p. 165.
56   Robert Hogarth Patterson, *Gas Purification in London, Including a Complete Solution of the Sulphur Question*, London: Blackwood, 1873.

Finally, if sudden darkness did not kill any French monarch, gas was indirectly responsible for some hundreds of deaths. In 1858, fifteen people were trampled to death at the Victoria Theatre in London during a panic caused by a small gas explosion.[57] Much worse, an explosion of gas in 1881 at the Nice opera house set fire to the scenery, and in the darkness, a terrible stampede occurred that blocked the exits. More than 200 people died from being crushed, asphyxiated or burnt.[58] That same year, a fire broke out at the Ring Theatre in Vienna, and the gas was turned off for safety reasons, but with deadly consequences: 'The passages were narrow and intricate, in the darkness they became choked with people rushing hither and thither, trampling and crushing one another to death in the mad struggle for life.'[59] A total of 609 corpses were found. With hindsight, Nodier and Pichot's warnings about panic inside a darkened theatre seemed prophetic. After the Nice disaster, oil lamps made a comeback in French theatres to guide spectators towards the exit in the event of the gas being cut off.

Even more accidents exceeded all the expectations of the opponents or experts. Indeed, the technical system created entirely unforeseeable effects. For example, until the 1880s, certain gas explosions remained inexplicable, for they took place without there being any flame. It came as a surprise to discover that poorly purified gas could contain acetylene, which reacted with the copper in the pipes to form copper acetylide, which would simply detonate on impact.[60] The copper pipes had to be banned and the presence of acetylene in the gas had to be monitored.

Real-world developments could also change how technology behaved. For example, in Strasbourg, in January 1841, extreme cold made the gas much drier than it otherwise would have been. Instead of depositing water in the siphons placed for this purpose at low points in the network, it dried them out. Since the ground was frozen, the gas could not escape into the open air and was stored in the cellars of the houses. One family died of poisoning. Even the most astute or the most twisted opponent of gas would have found it hard to imagine such a chain of circumstances.[61]

Technologies of different generations could also interact in unpredictable ways. On 12 July 1883, explosions nearly brought down several

57 *The Times*, 28 December 1858.
58 *Le Figaro*, 25 March 1881.
59 *The Times*, 10 December 1881.
60 Alexandre Layet, 'Le gaz d'éclairage devant l'hygiène', *Revue d'hygiène et de police sanitaire*, 2, 1880, p. 951.
61 Gabriel Tourdes, *Relation médicale des asphyxies occasionnées à Strasbourg par le gaz d'éclairage*, Paris: Baillière, 1841, p. 78.

buildings on rue François-Miron in Paris. A café was destroyed, and 86 people killed. Once again, the causes were complex, with the Compagnie parisienne d'éclairage and the Administration des eaux de Paris each blaming the other. The accident was linked to the intersection of several technical networks: a water leak had created a cavity, and a nearby gas pipe had lost its support, collapsed and broken. Since the cobblestones had been replaced by tarmac, the gas was unable to escape and accumulated in the newly built sewers before causing devastating explosions.[62]

7. The social construction of safety

Critics' gloomy predictions went unheeded – and most of the accidents that took place were unforeseeable, anyway. So, should we conclude that the controversy of the 1820s was just so much wasted ink? Quite the opposite. A study of the technological trajectory that gas has followed shows that the dispute played a major role in making it safer.

First of all, Pauwels, who was at the centre of public attention, was also the engineer who invented the most important safety mechanisms. His gigantic gasometer was the first to be fitted with a guide pin that prevented it from moving back and forth. For surveillance and safety reasons, his plant was built on a panopticon principle: the 336 distillation retorts were lined up in four parallel rows.[63] This spatial grid contrasted with the dark, cramped London gasworks, which the British select committee feared would amount to little surveillance and conditions conducive to sabotage.[64] Twenty years later, it was again Pauwels who designed the second generation of gasworks: a system of articulated pipes made the movement of the vessel more precise and regular, and the retorts were now made of cast iron rather than brick, and connected to the gasometer by an aspiration system that prevented the gases from escaping.[65] Yet surely his most important invention was the gas-compensator, patented in 1846, which regulated pressure in the gas grid. Friction, differences in altitude and the elasticity of the gas meant that pressure varied widely. Pressure fluctuations caused major leaks (in the 1840s, 25 per cent of gas

---

62   APP, DB 152.
63   Ibid., RCS, 4 October 1821.
64   'Report of the select committee on gas-light establishments', p. 72.
65   'Description des perfectionnements par M. Pauwels', *BSEIN*, January–March 1849.

evaporated into the atmosphere in Paris) and dangerous oscillations in the flame.[66]

The controversy was also at the root of the 1824 regulations, which, while authorising gigantic gasometers, also defined a minimum safety standard. The ways in which technological risk was managed in France and Britain were diametrically opposed. In France, it was obvious that a regulatory solution was needed. The question the government put to the *académicien*-experts was: What should the content of the decree be? In Britain, the parliamentary select committee asked: Is it really necessary to enact a law? And the experts' answer was unanimously negative: competition, progress and improved safety all went hand in hand. After all, did gas leaks not cost companies money? Regulatory intervention thus risked *undermining* the safety of the gas installations. The optimisation of technology, even from the point of view of safety, would be achieved through the market and competition. One expert, for example, regretted the intervention of the City of London, which had won a case against the Dorset Street plant. Under the terms of the judgement, the plant no longer had the right to discharge its washing water into the Thames. The company thus had to resort to solid lime, which turned out to be a less effective purifier than limewater. Through their ill-advised intervention, the authorities had thus unwittingly replaced a minor nuisance, which did no more than kill fish in the Thames, with a much more formidable risk: that the gas delivered to consumers was impure and thus more dangerous. The rejection of regulation was underpinned by faith in the inventiveness of entrepreneurs, subject to competition: the definition of a standard, fixing a technical norm in place, would ultimately be detrimental to progress in safety matters.

Although it is difficult to pinpoint the precise role played by the regulations in making gas lighting safer (which would require knowing what technical conditions would have been reached had the controversy not arisen), the appalling state of London's gasworks in the 1820s can provide an interesting point of comparison. Despite (or because) they were twenty years ahead of their time, despite the lower cost of materials, despite the presence of workers and engineers whose know-how was superior to that of their French counterparts, London's gasworks were at a much lower safety level than the first Parisian installations. At London's Whitechapel plant, for example, huge 15,000-cubic-foot canvas bags covered

---

66   Charles Combes, 'Sur un appareil imaginé par M. Pauwels', *Journal de l'éclairage au gaz*, 4, 1852.

with tar were used as gasometers. In the Brick Lane factory, secondhand brewers' vats served as gasholder tanks.[67] Raised tanks saved on the cost of digging a basin, but they often broke under the pressure of the water. Clegg also saw no problem in building gasometers in ponds to save on the construction of a reservoir! Indeed, he reported seeing nine of them floating, like water lilies, in Manchester.[68] Despite the relative cheapness of iron in England, wood was often used in all parts of the factories: raised tanks, gasometers, the roofs and frames of the sheds. Congreve was horrified by one factory where large quantities of coal tar were allowed to accumulate because the company did not want to pay the costs of removing this dangerously flammable residue. He compared this factory to an 'artificial volcano'.[69]

Generally speaking, it seems that the distribution of gas in London was affected by rudimentary production techniques. In the 1830s, companies advised users to keep a constant eye on their gas spouts, whose flame could be a foot high or go out suddenly, which was particularly dangerous when the gas supply resumed. The gas was so sulphurous that some individuals preferred to instal their spouts outside, even if it meant adding a reflector.[70] From the 1850s onwards, there was an abundance of editorials, lawsuits, petitions and proposed legislation. This movement of local residents, consumers and hygienists, which the newspapers labelled the gas agitation, was strengthened after the explosion of the gasometer at the Three Elms factory in 1865. In 1872, a Board of Gas was finally established in London to monitor plant safety and gas quality.

The irony of history is that in France, even though the industry was in its infancy and expertise was lacking, the social and administrative mobilisation for safety had already taken place half-a-century earlier, through a combination of public controversy, the importation of British expertise and the government's readiness to regulate the industry.

The history of gas tells us many things. First, it tells us about the existence of a form of reflexivity about the consequences of industrialisation. The Parisians of the 1820s questioned the many transformations that gas implied for their autonomy, their leisure, their safety, the air, the night, the beauty of the world, coal reserves and the economy. For Nodier and Pichot, the main challenge of the time was no longer the mastery of nature,

---

67   'Report of the select committee on gas-light establishments', p. 72.
68   Ibid., p. 52.
69   *Repertory of Arts and manufactures*, 43, 1823, p. 80.
70   *Annales de l'industrie nationale et étrangère*, 10, 1823, p. 319.

but the mastery of that mastery itself: 'Everywhere in society, man has acquired a greater development of his power with new instruments . . . what he has to avoid now is overestimating his power, abusing it, and turning it against himself.'[71]

Second, the history of gas demonstrates the necessity of opening up expert procedures to a wider variety of actors and competencies. Expertise on gas in France was monopolised by a small clique of chemists and engineers who hoped that they could help France catch up through their mathematical virtuosity. Taking such an approach, their expertise was akin to a thought experiment based on an ideal gasometer whose behaviour could be calculated according to the laws of physics. Disagreements were suppressed thanks to a procedure in the Académie that produced science as a single voice speaking the truth about the nature of the world. In contrast, the House of Commons' expert appraisal involved a wide variety of players. Engineers and foremen who handled the gasometers on a daily basis could testify to the problems they encountered. When experts expressed doubts or contradicted each other, the select committee strove to grasp the reasons for this uncertainty. Clearly, these contrasting expert practices led to results of unequal quality.

Third, the history of gas is emblematic of the radical unpredictability of technology. The consequences of gas lighting are still with us. Even though manufactured gas has disappeared from our cities, they have contaminated the soils with tars that contemporary medicine recognises as carcinogenic. The other surprise is climatic. In the late eighteenth and early nineteenth century, coal was often presented as a 'green technology' because it would reduce the consumption of wood and thus preserve forests. This argument is explicitly stated in the 1823 report of the Académie des sciences: gas would protect France from deforestation and hence counter climate degradation. Paris, which consumed 100,000 tonnes of wood a year in the 1820s, burned just 70,000 tonnes twenty years later. Meanwhile, coal consumption had risen from 50,000 to 180,000 tonnes, almost a third of it devoted to the capital's gasworks.[72] By appearing to free energy from the constraints of the forest, gas disinhibited energy consumption and contributed in no small measures to our climate predicament.

---

71   Nodier and Pichot, *Essai critique sur le gaz hydrogène*, pp. 1–2.

72   *Le Journal des économistes*, 6, 1843, p. 19. In Great Britain, at the end of the nineteenth century, 10 million out of a total 30 million tonnes were used in the gas industry. See Peter Thorsheim, *Inventing Pollution: Coal, Smoke, and Culture in Britain since 1800*, Athens, OH: Ohio University Press.

There is no point in blaming the experts of the 1820s for failing to foresee these dangers. But a historical comparison between what they predicted and what actually happened makes it possible to appreciate what the word 'uncertainty' means, when we use it in a detached way to talk about contemporary technologies. Technology and its circumstances – in other words, the complexity of the world – systematically thwarted all predictions. The only winner of the dispute was the unpredictable itself.

6

The Mechanics of Fault

Technical safety standards responded to two distinct political requirements. One, as we saw with gas lighting, was to protect industrial capital by circumscribing risk and setting it on a stable legal footing. The second concern, which will be the focus of this chapter, was to integrate the new products of the Industrial Revolution into a liberal legal anthropology based on the distinction between the responsible individual and the passive object.

The French Civil Code of 1804 resulted from a project for less government: the legislator sought to constitute society as a set of individuals whose behaviour would be harmonised through their judicial interactions. In this context, accidents were conceived as private matters involving a responsible party and a victim. These were wrongs subject to the law and had to be combated as such through the obligation to provide compensation. Following a long legal tradition, the quasi-tort liability defined by Article 1382 of the Civil Code made compensation conditional on the existence of fault. But, for this self-regulating system to work, it was still necessary to be able to identify fault, that is, to attribute human causes to each accident. This required being able to distinguish clearly between two ontological orders: that of individuals who could be apportioned blame, and that of passive objects.[1]

As far back as the eighteenth century, there was no reason to consider this distinction self-evident, and rules of justice had been devised to settle tricky cases. When objects had violent effects, the law generally laid the blame with their owners. According to Domat: 'The order that

---

1   See Paul Ricœur, 'Le concept de responsabilité. Essai d'analyse sémantique', *Le Juste*, Paris: Seuil, 1995, vol. 1, pp. 41–70.

binds men together in society . . . obliges each man to keep everything he possesses in such a state that no one receives either harm or damage from it.[2] In many circumstances, the fault was so slight that it was only presumptive: an animal's owner was always held liable for any damage it caused, as was the owner of a forest whose rabbits caused damage to neighbouring fields, or the builder who used a machine hoist to lift materials. The distinction between 'manslaughter through negligence' and 'accidental death' was, indeed, a legal construct: if a barber in his shop was jostled and he slit his customer's throat, he was innocent, but if the same accident occurred when he was working outdoors, he could be considered guilty of imprudent conduct.[3]

From the 1820s onwards, the technologies of the Industrial Revolution blurred the criteria for apportioning blame even more. Instead of one cause which implicated a person who could be held responsible for misusing things, judges and engineers had to deal with blurrier combinations of causes – indiscriminately mixing together error, inattentiveness, ignorance, unforeseeable technical malfunctions, wear and tear, material fragility, conditions of use and maintenance, and so on. In such cases, the cause of the accident was spread across a whole jumble of people and things, and this made it impossible to apportion blame and thus to set compensation. This symmetry between humans and non-humans, which the contemporary sociology of science regards as an investigative finding, was, for the legislator, a starting point and a problem to confront.[4] Putting things and people on the same level of accountability did not resolve any practical question of justice; and to accept that violence could arise from objects, not humans alone, meant depriving society of a powerful means of self-discipline; namely, the constant fear of wrongdoing and being punished for it.

In this context, technical standards played a fundamental political role. By producing (including in the sense of 'staging', as in a theatre) predictable objects that could not cause accidents all by themselves, the standard made it possible to systematically direct blame towards humans. By confecting the notion of a perfect technology – indeed, whose perfection was

---

2 Jean Domat, *Les Loix civiles dans leur ordre naturel*, Paris: Coignard, 1691, vol. 2, pp. 113–37.

3 Daniel Jousse, *Traité de la justice criminelle*, Paris: Debure, 1771, pp. 519–27.

4 Michel Callon, 'Éléments pour une sociologie de la traduction. La domestication des coquilles Saint-Jacques dans la baie de Saint-Brieuc', *L'Année sociologique*, 36, 1986; Diane Vaughan, *The Challenger Launch Decision: Risky Technology, Culture, and Deviance at NASA*, Chicago: University of Chicago Press, 1996; Peter Galison, 'An accident of history', *Atmospheric Flight in the Twentieth Century*, Dordrecht: Kluwer, 2000, pp. 3–44.

guaranteed by the state administration – the standard aimed to produce human subjects who could be held responsible for accidents.

Many historians have analysed the transformations that took place through the ways that societies managed accidents at the end of the nineteenth century. Legislation on workplace accidents (1884 in Germany, 1898 in France) introduced a system of no-fault liability: workers no longer had to prove to the courts that their employer was at fault before they could secure compensation. As François Ewald explains, accidents were removed from the logic of the justice system and subjected to the different logic of occupational risk; they were no longer the result of human fault, but the inevitable downside of the industrial production from which all society benefited. If these were the inherent costs of progress, it was only right that society as a whole should foot the bill, through insurance mechanisms. In this reading, making the law 'social' consisted in its adaptation to the development of the world of production and its mechanisation.

The problem with this account is that it amounts to presenting the socialisation of professional risk (in France, meaning the 1898 law) as if it created a regulation *ex nihilo* and it imagines that the prior situation was just pure laissez-faire at the expense of the worker. In so doing, it overlooks the vast legal and technical efforts that had gone into preserving the principle of liability throughout the nineteenth century. It is odd to note, in this regard, that liability for fault was erected as a fundamental principle of civil law at precisely the moment when the technologies of the Industrial Revolution made its application into a source of injustice. What requires historical explanation, then, is not only the recognition of occupational risk that came at the end of the nineteenth century but also the legal application of the principle of liability that preceded this point.[5]

This is especially true given that France's Civil Code tolerated a degree of flexibility. While Articles 1382 and 1383 emphasised the link between fault and reparation, the ones that followed acknowledged instances of liability in which fault was only presumed. Hence, an animal's owner was liable for any damage it caused. Accidents in the army, in the civilian navy and in public works led to the payment of a pension without any need to prove fault.[6] Finally, the Civil Code introduced liability for

---

5   François Ewald, *Histoire de l'État providence. Les origines de la solidarité*, Paris: Grasset, 1986. Another problem is that fault remained essential to civil law and in industrial life itself. See Francis Chateaureynaud, *La Faute professionnelle. Une sociologie des conflits de responsabilité*, Paris: Métailié, 1991.

6   Philippe-Jean Hesse, 'Les accidents du travail et l'idée de responsabilité civile au XIXe siècle', *Histoire des accidents du travail*, 6, 1979, pp. 1–57.

damage caused by 'things in one's custody'. The courts could have equated machines with the horses they replaced or invoked the 'custody of things' to make the owners pay. This was, indeed, what they did, but not until the end of the 1860s. Before then, compensation was only awarded in cases of proven fault. Accidents with undetermined causes were regarded as chance events, that is, on the same level as natural disasters, and did not imply any compensation.

This chapter, then, has two goals. The first is to identify, through a focus on steam engines, the material infrastructure of legal liberalism; that is, the technical devices that made it possible to maintain the simulacrum of a liable individual. The second is to show that the theories of occupational risk and compulsory insurance were far from a break with liberal economic principles, but, instead, followed in line with Chaptal's old project of providing security to industrial capital.

1. Standards, responsibility and self-discipline

Liability is based on predictability: an accident is grounds for compensation insofar as the judge can apportion blame, that is, show that an individual failed to *foresee* the consequences of his own action. To produce responsible human beings, technology thus had to be made as predictable as the laws of nature. At first sight, in the technological context of the 1820s, this ought to have posed no difficulty. Steam obeys linear mathematical laws ($PV = nRT$) and seems emblematic of a predictable and controllable technical system.

Yet, throughout the century, steam would, in fact, pose irresolvable problems. While most explosions could be explained by an overloaded valve or damaged sheet metal, others remained complete mysteries. For example, boilers sometimes exploded when the valve opened. The explanation ventured by the engineers was the following: the metal sheet built up heat and the water kept under pressure did not boil. When the valve opened, the pressure dropped and the boiling water hit the overheated metal and evaporated, thus causing the explosion. The theory of the spheroidal state of water explained the explosions that occurred when the boiler cooled down: the water suspended on a cushion of superheated steam suddenly fell onto the burning-hot metal sheets. Other hypotheses cited the formation of hydrogen gas upon the water's contact with the metal, a delay in boiling due to the lack of gas in the water, or even electrical phenomena. In the technical literature of the time, there were so many theories on sudden explosions that they made up a genre unto themselves.[7]

If the 'age of steam' exalted progress, this was certainly not because it was unaware of its possible dangers. When high-pressure boilers (at above twice atmospheric pressure) appeared in France in 1815, industrialists had, in fact, hesitated to use such unpredictable devices, particularly in shipping. In 1822, the Société d'Encouragement pour l'Industrie Nationale published a glowing report on the *Zoolique*, a boat built in Nantes which used paddle wheels but replaced the steam engine with a horse.

In March 1822, after a series of explosions in the United States and Great Britain were reported in the Paris press, Interior Minister Jacques-Joseph Corbière banned high-pressure boilers within seventy metres of dwellings, in effect banning their use in towns. The Conseil de salubrité, this ardent defender of industrial interests, refused to apply this decision. In its view, since it was enough to overload the safety valve to transform a low-pressure boiler into a high-pressure one, this measure threatened to become an effectively total ban on steam. In any case, high-pressure boilers were much more efficient, at least twice as much as standard boilers. The savings in coal consumption were considerable, as was the reduction of the pollution connected to its combustion. The Conseil de salubrité openly criticised the minister's decision for not being 'justified by a properly considered discussion of the principles of the matter'.[8]

Steam posed a twofold problem in policy terms. First, increasing pressure also heightened the power of the engine, together with the pro-duction rate and thus output. The pursuit of profit obviously made for greater risks. Second, the main danger did not owe to a faulty arrangement of the technology but to its misuse. In July 1822, the Conseil de salubrité authorised a boiler in exchange for a *commitment* from the proprietor not to exceed two atmospheres of pressure. The prefect of police rejected this compromise: 'Safety must be achieved in the construction of the device itself.'[9]

Faced with the Conseil de salubrité's criticisms, the interior minister cancelled its decision and commissioned a report from the Académie des sciences. In April 1823, Laplace, Girard, Dupin, Prony and Ampère

---

7   Jacob Perkins, 'Sur les causes des explosions les plus dangereuses de machine à vapeur', *Annales annuelles de l'industrie manufacturière*, 2, 1827, p. 286; Pierre-Hypolite Boutigny, Études sur les corps à l'état sphéroïdal, Paris: Librairie scientifique, 1847; François Arago, 'Explosions des machines à vapeur', Œuvres complètes, Paris: Baudry, 1855, vol. 5, pp. 118–80.

8   APP, RCS, 23 August and 18 September 1822.

9   APP, RCS, 12 July 1822.

Out of reach valve, early nineteenth century[10]

proposed a technical solution to the problem of monitoring the equipment. Two safety devices were added: a second valve 'placed so as to remain out of reach of the worker' and two 'self-fusing plates' made of an alloy which would melt at a temperature corresponding to the maximum pressure.[11] In each case, the aim was to restrict the freedom of the engineman, who was assumed to be the cause of the accidents.

In October 1823, an ordinance stipulated the use of these two safety devices and set boilers under the control of the mining administration. A central steam engine commission was set up under the auspices of the Interior Ministry. In the *départements*, mining engineers were required to check that boilers complied with the regulations, test them for pressure, inspect them once a year and affix official stamps to the valves and fusible washers.[12]

Such administrative controls immediately posed a new technical problem: for to set the melting points of the washers, it was necessary to know what law there was linking temperature to pressure. In 1823, the government asked the Académie to establish this relationship up to the considerable pressure of twenty-four atmospheres. The scientists involved in this experiment (Arago, Prony, Ampère, Girard and Dulong) did not settle for approximations. For the sake of greater accuracy, they

---

10   Deutsches Museum, Munich.

11   *Procès-verbaux des séances de l'Académie*, vol. 7, 14 April 1823.

12   These proofs were essential for authorisation. Steam engines were classified in the second class of the 1810 decree.

abandoned the use of a spring manometer and instead installed a twenty-metre glass tube along the tower of Paris's Collège Henri IV. According to Arago, this was one of the largest physics experiments ever conducted.[13] Thanks to the precision and courage of the *académiciens*, French technology based on mathematical laws would distinguish itself from British or American empiricism. The new industrial devices and the government's determination to make them safe allowed the physics experiment to reach an unprecedented scale. Two decades later, in 1836, the great experimenter Victor Regnault obtained considerable funding from the Ministry of Public Works to accurately determine the laws of steam at very high pressure. He took advantage of this to conduct thermometrical research into delays in boiling and fixed points.[14]

Unlike pressure, the resistance of metal plate was difficult to measure, since it varied according to the quality of the materials used and the temperature. Since no clear mathematical law could be arrived at, the *académiciens* delegated the work to simple engineers. Because of this uncertainty, the administration was forced to impose a considerable margin of safety: the thickness of boilers was calculated to withstand a pressure twice the test level, which was itself set at five times the working pressure, giving a safety margin of ten.[15] Boiler manufacturers were outraged by the waste of metal and extra expanse which this involved. British engineers were also sceptical about what point there was to all this. In their view, safety had more to do with the quality of the sheet metal used by the boilermaker. Although there was no official standard in Britain, in the 1850s the Manchester Steam Users Association recommended a safety factor of five.

In fact, the safety margin was essential to the standardisation project. The administration that sought to prevent accidents 'upstream' through technical prescriptions also had to anticipate *the wear and tear* that

---

13   *Mémoires de l'Académie des sciences de l'Institut de France*, Paris: Gauthier-Villars, 1831, vol. 10, p. 235. Philadelphia's Franklin Institute used a very different method. The problem of explosions was not considered as a linear phenomenon – in terms of the battle between pressure and the resistance of the materials – but as an unforeseeable occurrence. Rather than seek to establish a physical law, the American engineers tried to reproduce the mysterious explosions cited by mechanics. They built a boiler fitted out with a thick sheet of glass to let them observe what was going on inside. See *Report of the Committee of the Franklin Institute of the State of Pennsylvania on the Explosions of Steam Boilers*, Philadelphia, 1836.

14   Victor Regnault, *Relation des expériences pour déterminer les principales lois et les données numériques qui entrent dans le calcul des machines à vapeur*, Paris: Firmin Didot, 1847, pp. 2–5.

15   'Troisième instruction relative à l'exécution des ordonnances sur les machines à vapeur', *Annales des mines*, second series, 3, 1828, pp. 490–513.

Explosion in a Paris workshop, 1867 – five dead[16]

would weaken the metal. This was how mining engineers justified the extra expense: 'Manufacturers will adjust to these thicknesses and only sell boilers which will still be able to withstand the test pressure *despite their long use*.'[17]

In 1828, the essential parameters of French boilers were standardised: the thickness of the sheet metal, the melting point of the fusible washers and the diameter of the valves could be calculated from the working pressure and the diameter. This was probably the first time that equations had defined the legal form of a technical object.

What happened when one of these French boilers exploded? What effect did standards, safety margin and administrative certification have on the way liability was apportioned?

Determining the cause of an explosion was a difficult task: the boiler attendant was often killed, and the workshop devastated. The *département*-level mining engineer would look for clues, collect witness statements and analyse samples of the metal. The opaque circumstances of the accident could have left the question wide-open to interpretation – that is, if it had

---

16   AN, F14 4217.
17   Ibid., p. 516.

not been for the four pre-established theories recognised by the Corps des Mines, which inevitably framed his report. These four were the overloading of the safety valve, the lowering of the water level, construction defects, and the wear and tear of the sheet metal. These ready-made explanations provided standard narratives for accident reports and made it possible to set clear boundaries to the possible chains of cause and effect.

This had two consequences. First, by setting out the causes and attributing these explanations with scientific legitimacy, the Mining Administration provided judges with legally responsible individuals. Between 1827 and 1848, out of fifty-eight boiler explosions, only four remained unexplained.[18] The judiciary itself was generally more cautious and prosecutors were reluctant to prosecute based on engineers' reports. For them, a mere failure to comply with standards was insufficient grounds for a conviction.[19]

Second, the mining administration often blamed accidents on workers: fully one-third of explosions were said to be a result of their errors, inattention or carelessness (failing to keep up the supply or overloading the valve). The remaining two-thirds were put down to construction or maintenance faults attributable to the owners. Yet, in the 1850s, insurance companies in Britain held workers to be at fault for only one in ten explosions.[20] What can explain such a discrepancy? In France, when a boiler exploded, the mining engineer was blamed for his inability to monitor the boilers in his *département* and keep them in good condition. In his report to his superiors in Paris, it was thus in his interest to find fault with the worker, rather than the poor condition in which the boiler had been kept. Moreover, the central commission on boilers sometimes rejected conclusions blaming the worker and criticised the mining engineer's overly convoluted explanations.[21]

The engineers' reports similarly focused on the defects of the engine and those of the worker. It was conveniently discovered that, at the time of the accident, the engineman was drunk and had fallen asleep – or was

---

18   'Tableau du nombre des accidents survenus dans l'emploi des appareils à vapeur', *Annales des mines*, 15, 1849, pp. 22–45.

19   AN F14 4215, explosion in 1830 in Elbeuf, AN F14 4217, explosion in 1868 in Reims. In 1827, the Haut-Rhin mining engineer reported that of the twenty boilers in his department, only four were fitted with fusible washers (AD Haut-Rhin, 5 M 47, Rapport de l'ingénieur sur les machines à haute pression).

20   Association for the Prevention of Steam Boiler Explosions, *Reports of the Proceedings*, 1864.

21   AN F14 4217, explosion in 1867 in Avignon.

even leading the 'dissolute life'![22] Guidelines from 1824 explain that the boiler attendant must be 'not only attentive, active, clean and sober, but also free from any defect that could harm the regularity of the service. Nothing must disturb his attention during work; otherwise there can be no safety in the establishment.' As for valve overloads, the instructions warned that 'they are extremely dangerous . . . workers must be aware that one of the main effects of an explosion would be to release a huge quantity of burning steam which would cause them a cruel death.'[23] The danger incurred by the worker encouraged discipline and thus improved safety. If need be, a perfectly predictable steam engine could also be a good punishment mechanism.

2. Vices and improvements

To the engineers' minds, if the worker was not responsible for an accident, then such an outcome must reflect some *ephemeral* condition of the technology itself. Two concepts used in this regard – 'construction defects' and 'perfection' – set out the perspective of a simpler world made up of perfect machines and responsible humans.

The focus on 'construction flaws' presupposed the possibility of distinguishing the essence of the technology from its defective real-world realisations. For example, when a boiler exploded, if it was made of cast iron the engineers would blame the material; if it was made of wrought iron plates riveted together (as per their advice) they would blame poor construction – perhaps the rivets were too close together, too far apart, too aligned, too deep, oval, not bevelled enough, and so on. There were a thousand and one ways to blame a rivet rather than the intrinsic dangers of steam power. For the mining engineers, the construction defect had the advantage that it was hidden: since it could only be discovered once the explosion actually occurred, it did not raise any doubts over the diligence of their monitoring work.

Reassurance was also to be found in the key notion of 'perfectionnement', as it implied the existence of a perfect state of technology. The competition launched in 1829 by the Société d'Encouragement pour

---

22   Charles Combes, 'Sur les explosions de chaudières à vapeur depuis 1827', *Annales des mines*, second series, 20, 1841, pp. 130–9.

23   *Instruction sur les mesures habituelles à observer dans l'emploi des machines à vapeur à haute pression*, 19 April 1824.

l'Industrie Nationale put it in explicit terms: two prizes were to be awarded to anyone who *'perfects* and *completes* the means of safety' or found 'a boiler shape and construction that prevents or *erases any danger of explosion'*.

In the 1830s, perfecting the machine, above all, meant disciplining the worker: for every cause of error or negligence, a mechanism had to be found to control or replace the boilerman's role. Various valve models were proposed, refining the devise stipulated in the 1823 decree: valves inside the boiler and thus inaccessible to the worker, valves with enclosed bearings and valves with exhaust levers. Other devices were designed to warn the worker of the danger or raise the alarm over his own inattention: for example, manometers striking a bell when the pressure was too high, or floats that allowed a jet of steam to pass through a whistle when the water level was too low.[24]

Other inventors instead sought to replace the worker with automatic mechanisms: a float which, when lowered, opened the water supply valve; a gauge which opened the valve if the pressure passed a certain threshold. In the mid-nineteenth century, explosion prevention was one of the key areas of development for the automation and interconnection of systems. Inventors tried to establish negative feedback loops by connecting the boiler's different parameters: how was it possible to glean information on pressure and link it to the furnace's air supply? How about measuring the water level, and adjusting the tap accordingly? With the mechanical means available in the nineteenth century, this was no simple task.

The notion of guaranteeing safety through automation – and indeed, by bypassing the workers' role – gave material expression to the productive utopia advanced by the eighteenth-century philosophers. The idea was to subordinate the worker to the knowledge and creations of the engineers. In 1767, the philosopher Adam Ferguson wrote about factories, explaining that their 'perfection consists in being able to do without the mind ... so that ... the workshop can be considered as a machine whose parts are men'.[25] Fifty years later, automation for safety purposes was still interpreted in these terms. On the subject of steam engines, a French encyclopaedia marvelled: 'These refinements ... complete the task of turning this apparatus into a kind of automaton, capable, so to speak, of executing

---

24 'Rapport sur le concours relatif aux moyens de sûreté contre les explosions', *BSEIN*, 1832, pp. 452–70, and *BSEIN*, 1833, pp. 108, 300.

25 *Histoire de la société civile*, Paris: Desaint, 1783, vol. 2, p. 109; Simon Schaffer, 'Enlightened automata', *The Sciences in Enlightened Europe*, Chicago: University of Chicago Press, pp. 126–65.

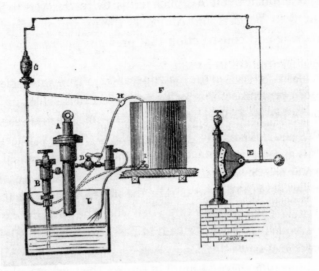

Thibaut automatic power supply[26]

alone and without supervision the service demanded of it.'[27] The London physician Herbert Mayo ventured to compare the metabolism of humans with that of machines, for the latter are also self-regulating organisms: 'In its present perfect state, the steam-engine . . . regulates the quantity of steam admitted to work, the briskness of the fire, the supply of water . . . it opens and closes its valves with absolute precision . . . and when anything goes wrong, it warns its attendants.'[28] In the *Grundrisse*, Karl Marx was also interested in the mechanisation of man and the humanisation of mechanics, but this reduction of differences, far from being progress, seemed to him to be symptomatic of the disorders of capitalism, which transformed the worker 'into a mere living accessory of the machine.'[29]

In the 1820s, the worker was seen as a more irregular creature than the machine, and more difficult to perfect. It thus seemed that safety could be achieved by embedding human actions in the technology itself. In the nineteenth century, this equating of man and machine, the exchange between their characteristics and capacities to act, was the dominant way

---

26   *Les Mondes. Revue hebdomadaire des sciences*, Paris: Rothschild, 1866, vol. 10, p. 539.

27   *Encyclopédie moderne, ou dictionnaire abrégé des hommes et des choses*, Brussels: Lejeune, 1832, vol. 23, p. 205.

28   Herbert Mayo, *The Philosophy of Living*, London: Parker, 1838, p. 21.

29   Cited in André Gorz, *Métamorphoses du travail. Critique de la raison économique*, Paris: Galilée, 1988, p. 94 [English translation of Marx from *Marx's Grundrisse*, London: Palgrave Macmillan, 1980, p. 143].

of thinking about technology and of integrating it into a liberal social order.

3. The liability market in Britain

Insurance and liability are often considered as two antithetical ways of managing accidents. Whereas the first is a collective technology based on statistics and anticipation, the second is individual in character, and manages situations *a posteriori*. The social impact of these two techniques would also appear to differ considerably: on one side, a population of premium-payers and beneficiaries; on the other, a legal confrontation between the purported responsible party and the victim. These counter-positions – modelled on the discourse put forward by the 1890s promoters of the doctrine of occupation risk – tend to reduce insurance practices to the principle of insurance. But, if we look more closely at the realities of British steam-engine insurance in the 1850s, this opposition between two different worlds is put into a more relative perspective. Far from separating accidents from the liability regime, such insurance instead tended to impose the categories of cause and fault on a reticent legal system. Insurance companies' aim was to sell policies to industrialists. In order to do so, they had to make the entrepreneurs liable, thus encouraging them to take out the policies which they had to offer.

Between 1830 and 1846, in order to compensate the victims of technological accidents even without liable individuals, English juries turned to the 'deodand' – a medieval legal device that allowed the authorities to confiscate an object (or animal) that had injured or killed someone.[30] The forfeited object was sold, and the money collected was used to compensate the victims. With 'deodand', the owner could be punished through no fault of his own, through the attribution of a quasi-intention to the possessions deemed guilty. Challenging the distinction between person and thing, deodand already seemed archaic to eighteenth-century English jurists. Yet, from 1830 onwards, English juries faced with technological accidents used this system to compensate victims of boiler explosions or railway accidents.[31]

---

30   From *Deo dandum*: 'That which must be consecrated to God'. 'Deodand' could only be applied to moving or animate things: horses, cattle, stagecoaches, mill wheels, and so on.

31   Harry Smith, 'From Deodand to Dependency', *The American Journal of Legal History*, 11, 1967, pp. 389–403; Elisabeth Cawthon, 'New Life for the Deodand: Coroners' Inquests and Occupational Deaths in England, 1830–1846', *American Journal of Legal History*, 33, 1989, pp. 137–47.

To understand this resurgence of this archaic legal oddity, we need to compare how British and French juries respectively dealt with accidents. In Great Britain, there was no administrative regulation of boilers. Parliamentary enquiries followed one after another (1817, 1819, 1832, 1844, 1870), but the first regulations on fixed boilers only came in 1902.[32] In the absence of any legal standard to specify boilers' proper technical form, the jury had to assess the owner's responsibilities with regard to their acceptable uses or correct maintenance. Deodand was linked to the difficulty of defining a cause of explosion and a responsibility. To take one example, in 1838, the boiler on the steamship *Victoria* exploded in London.[33] The jury, mainly made up of shopkeepers, questioned steam boiler mechanics and manufacturers. There were as many explanations as there were witnesses. The manufacturer emphasised the quality of the boiler, while his competitors blamed the shoddy materials used; a scientist claimed that the explosion owed to the formation of hydrogen gas; the survivors suspected overloading; and so on. The investigation failed to find fault, and the jury was unsure which verdict to return.[34] In France, in contrast, the mining engineer would reconstruct *one* story of the explosion that prevailed over competing versions of events, and which delivered a cause and a culprit. The French judge could also base guilt on a breach of standards. English juries had to resort to the more fluid notions of 'lack of care', 'reasonable precaution' or 'common knowledge'. In this world of blurred liabilities, they could not try owners for manslaughter and generally settled for passing a verdict of accidental death while imposing a deodand on the owner.

For around fifteen years, this medieval device for punishing guilty objects allowed industrialists to pay for the consequences of modern technologies. English juries sometimes awarded considerable deodands: £1,500 for the explosion of the steamship *Victoria*, and £2,000 for a railway accident in 1841. Deodands became a sufficiently mighty legal weapon to threaten small railway and steamship companies. Juries were accused of putting capitalists on the back foot and, in 1846, the British Parliament finally abolished deodands under pressure from the rail firms.[35]

---

32   Peter Bartrip, 'The state and the steam boiler in nineteenth-century Britain', *International Review of Social History*, 25, 1980, pp. 77–105.

33   'Victoria explosion inquest', *The Mechanics' Magazine*, 29, 1838, pp. 340–68.

34   'Explosions of steam boilers', *The Times*, 23 August 1838.

35   Peter Bartrip and Sandra Burman, *The Wounded Soldiers of Industry: Industrial Compensation Policy, 1833–1897*, Oxford: Oxford University Press, 1983, pp. 97–102. To compensate, the Campbell's Act, which abolished the legal principle 'the action dies with the person', enabled the family of a person killed in an accident to secure compensation.

In 1854, faced with mounting numbers of explosions, a group of Manchester industrialists led by prominent engineer William Fairbairn founded the Manchester Steam Users' Association. Fairbairn believed that industrialists should deal with the problem of accidents themselves, rather than risk government intervention. After the Factory Acts of 1844 (which imposed protections on the moving parts of machinery) and the 1852 Ordinance on marine steam boilers, the Home Office threatened to standardise industrial boilers. But it was feared that French-style norms would increase the cost of machinery and hobble technical innovation.[36] Fairbairn also rejected the insurance system on the grounds that the prospect of payouts would encourage owners to let up their vigilance. The Steam Users' Association thus offered its members five boiler inspections a year from an engineer and advice on how to reduce their coal consumption. The aim was to reduce risk to zero, and therefore not have to compensate for it. But the association was a failure: in the 1860s, the number of boilers under its control stagnated at below 2,000.

In 1859, the association's chief engineer Longridge defected to found the Steam Boiler Insurance Company with a London insurer. This was the first insurance against steam engine *explosions*, as the traditional insurance companies (Norwich, Phoenix, Imperial, and so on) only covered the risk of *fire* associated with boilers.[37] The Steam Boiler Insurance Company was highly profitable. Unlike life insurance or fire insurance, which covered cumulative risks (epidemics or urban fires), boiler explosion risks were independent. Steam engine insurance thus required a small amount of capital (£20,000 in the case of the Steam Boiler Insurance Company). By 1865, the firm was insuring more than 10,000 boilers at a cost of around £1 a year. It paid annual dividends of 20 per cent of its capital. Against this backdrop, competitors were founded: in 1862, the Midland Steam Boiler Inspection and Assurance Company, in 1864 the National Boiler Insurance Company, in 1865 the Steam Boiler Insurance Company, in 1873 the Yorkshire Boiler Insurance Company, in 1878 the Engine and Boiler Insurance Company and in 1882 the Scottish and

---

But the cost of this procedure and in particular the difficulty of proving fault made it much harder to obtain damages.

36   Robert H. Kargon, *Science in Victorian Manchester: Enterprise and Expertise*, Baltimore: John Hopkins University Press, 1977, pp. 41–8; William Fairbairn, *Governmental Boiler Inspection: Letter to John Hick MP*, 1870. According to another engineer, 'Fixed rules and routines check the spirit of enterprise': *The National Boiler Insurance Company, Chief Engineer Report*, Manchester, 1870.

37   Robin Pearson, *Insuring the Industrial Revolution, Fire Insurance in Great Britain, 1700–1850*, London: Ashgate, 2004, p. 181.

English Boiler Insurance Company. By the 1880s, half of Britain's 100,000 boilers were insured.

The British safety strategy based on insurance was very different from that of the French mining administration. Whereas the latter's aim was to prevent accidents by defining the proper technical form of a boiler, British insurance companies saw them as evolving objects; an explosion was not the result of faulty construction but of a deterioration process. Fairbairn, the great advocate of the inspection system, was also a pioneer in the study of material fatigue. His lectures on boiler construction testify to his concrete relationship with the materials, whose resistance he studied in function of their origin, the setting of rivets and the direction of the metal fibres.[38] British insurance companies fitted into this metallurgical and dynamic safety culture. Risk premiums were calculated based on a combination of two criteria: the category of boiler, which depended on its history (the quality of metal, nature of the feed water, riveting processes, the reputation of the manufacturer, the fixing of the boiler, previous repair work, and so on) and its age, which largely determined the variation in the premium (rated from one to four).[39] Unlike the French administration, which focused on the form of the technology, the insurance companies looked at the problem of boiler safety in terms of the object's lifespan.

Insurance inspectors had less interest in imposing technical forms (although they would sometimes ask for repairs before insuring the boiler) as in ensuring that structures were properly maintained. They generally conducted four inspections a year, including an internal one, as opposed to just one by the mining engineers in France. They examined how far corrosion had progressed and suggested repairs. The main aim of insurance was not to compensate for risk, but rather to cancel it out. For example, the Steam Boiler Insurance Company charged 17 shillings a year per boiler for inspection and only 3 shillings for insurance. In short, the insurance companies were doing the same job as the French mining authorities, but with far greater resources. We see this in the annual explosion rates: in contrast to the average six explosions per 10,000 boilers (in both Great Britain and France), the explosion rate for boilers subject to inspection by the insurance companies was 2 or 3 per 10,000.

Insurance also had a quite different effect on the allocation of liability than did the work of France's mining administration. Insurance

38 William Fairbairn, *Two Lectures on the Construction of Boilers and on Boiler Explosions, with the Means of Prevention, Delivered before the Leeds Mechanics' Institution*, 1851.
39 *Bulletin de la Société industrielle de Mulhouse*, 36, 1866, p. 403.

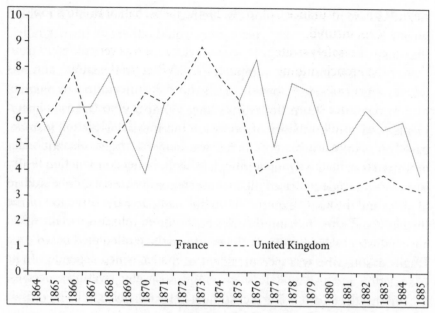

Number of explosions per 10,000 boilers[40]

companies' aim was to sell policies to industrialists. They thus had an interest in making them legally liable for explosions. Hence the extremely active communications aimed at coroners and legislators, denouncing the injustice done to the workers and pointing out the employer's responsibility for maintaining the boiler. Right from its foundation, the Manchester Steam Users' Association sent an engineer to the scene of each explosion so that he could study its causes and assist the coroner's investigation.[41] The director of the Steam Boiler Insurance Company made it perfectly

---

40   The graph clearly illustrates the effect of private inspection on boilers in Britain from 1875 onwards. In France, the inspection system did not develop until the 1880s. The British numbers carry some measure of uncertainty because, in the absence of administrative records, the total number of boilers is unknown. Insurers' estimates agree that the number of boilers rose from 90,000 in the 1860s to 120,000 in the 1880s. Sources: *Statistics Relating to Boiler Explosions*, Manchester, 1869; Cornelius Walford, 'The increasing number of deaths from explosions, with an examination of the causes', *Journal of the Society of Arts*, 29, 25 March 1881, pp. 399–414; Christine Chapuis, 'Risque et sécurité des machines à vapeur au XIXe siècle', *Culture technique*, 1982, no. 11, pp. 203–17. As compared to Europe, the risk in the United States was of a wholly different order of magnitude. In the 1880s, there were three times less explosions per capita in Great Britain than in the United States; more boilers exploded in one month in the United States than in a year in France or Germany. See Robert Thurston, *A Manual of Steam Boilers: Their Design, Construction, and Operation: For Technical Schools and Engineers*, New York: John Wiley, 1907, p. 717, and Louis C. Hunter, *Steam Power, A History of Industrial Power in the United States, 1780–1930*, Charlottesville: University Press of Virginia, 1986.

41   'Report from the select committee on steam boiler explosions', *House of Commons Papers, Report of Committees*, vol. 12, 1871, question 152.

clear: 'As long as boiler explosions are regarded as accidents for which no one is responsible, many steam users would rather run the risk rather than take out insurance.'[42]

According to insurance companies, in the large majority of cases, the worker was not liable.[43] Some insurers did their utmost to demonstrate this: in 1880, the Steam Boiler Insurance Company carried out experiments to disprove a theory of explosion incriminating the engineman, based on a faulty water supply. In fact water injected into a red-hot boiler produced tears in the wrought iron plate, without causing a violent explosion. Insurers also criticised the incompetence of coroners, the make-up of juries and the bogus experts whom the manufacturers hired to present the latest and often fanciful theories on sudden explosions.[44] The insurance industry lobbied energetically to reform the inquest process. In 1882, Hugh Mason, who was vice-president of the Manchester Steam Users' Association and also an MP, finally won his case by joining forces with the unions: the Steam Boiler Act stipulated that there must be an official engineer's report on each explosion.[45]

4. The technological crisis of human liability

In France, the 1840s saw a crisis in the technological regime for the regulation of accidents. This owed first of all to the fact that, within the world of manufacturing, the mechanical production of fault had in any case always been contested. Steam engineers had no confidence in the safety mechanisms imposed by the mining administration. The Société Industrielle de Mulhouse even refused to apply the 1823 decree because it considered it dangerous: locking a valve, for instance, made it impossible to check whether the disc was sticking to the boiler.[46] In 1830, a boiler attendant manual published by the École Centrale des Arts et Manufactures explained that the mining engineers 'knew little about the engines' and that their instructions were counter-productive: compliance with safety standards provided legal protection for owners, but most accidents in fact owed to poor maintenance of the machines for which they were

---

42   *Statistics Relating to Boiler Explosions*, Manchester, 1869, p. xx.
43   Manchester Steam Users' Association, *Chief Engineer's Monthly Report*, 1863.
44   Ibid., 1869.
45   Bartrip, 'The state and the steam boiler in nineteenth-century Britain'.
46   AD Haut-Rhin, 5 M 47, Isaac Schlumberger, Comité de mécanique, procès-verbal, 2 August 1827.

responsible. The author described 'the *care* required by steam engines, the *ailments* they suffer from, their *symptoms* and the remedies to be applied'. Mere compliance with standards was no guarantee of safety. Rather than monitor the engineman, it was better to train him in improved moral attitudes: 'The habit of cleanliness makes workers attentive, and by looking after a machine . . . that is well cleaned, they become attached to it: it becomes their property, as it were, a part of themselves, and they end up cherishing it as much out of affection and self-respect as out of duty.'[47] The worker's hygiene was thus matched by that of the boiler: an immaculate machine was the good mechanic's calling card.

At the same time, the qualities of a good worker were changing, and it was no longer enough to be 'sober', 'vigilant' or 'constant'. They now had to be capable of reasoning, initiative and even courage. In 1865, another manual stated that it was essential for a boiler attendant to have 'a great deal of composure and presence of mind'.[48] If the technical regulations of the 1820s had envisaged a disciplinary relationship between perfect technology and the faulty human worker, over the nineteenth century this was replaced by a complex web of affects (care, love, intelligence, initiative, courage).

From the perspective of the mechanics and enginemen – as far as this can be ascertained from primary sources – the problem of accidents could not be solved by technology, either. Some even denounced the absurd build-up of devices designed to control them. According to Adrien Chénot, a metalworker,

as the number of predicted causes of explosions increases, we will be able to add to the stamps, pressure gauges, thermometers, valves, washers, floats, gauged taps, plunger level tubes, distributors, whistles, etc., hygrometers, electrometers, capillarimeters . . . so that the top of the boilers would soon become like an académicien's table, cluttered with instruments exposed to all the blunders that result from confusion.[49]

A candidate in the competition on steam boilers organised by the *Société d'encouragement pour l'industrie nationale* proposed a remedy for explosions that was radically opposed to that of the engineers: not to protect

---

47   Philippe Grouvelle, *Guide du chauffeur et du propriétaire de machines à vapeur*, Paris: École centrale des arts et manufactures, 1830, pp. 11, 243.

48   A. Jaunez, *Manuel du chauffeur. Guide pratique à l'usage des mécaniciens, chauffeurs et propriétaires de machines à vapeur*, Paris: Hetzel, 1865.

49   Adrien Chénot, *Les Chaudières à vapeur sont des machines électriques. Les moyens de sûreté actuels sont impuissants*, Paris: Carilian-Goeury et Vor Dalmont, 1844.

the machine from human error but, conversely, to surround the machine with a metal shell (called a 'parapeur' – a 'fear-shield') that would protect the worker.

Workers wanted to preserve their autonomy from control by either technology, foremen or bosses. Only on this condition could they be held responsible for accidents. An engineman working on a steamship explained that explosions raised the question of who had authority over the ship. He was sometimes obliged, on the captain's orders, to carry out manoeuvres that he knew were dangerous. Since the engineman holds the passengers' fate in his hands, he must also be master on board and have full power to order repairs or scrap obsolete boilers. Once this reform was realised, explosions could be attributed to him.[50]

In *L'Atelier, organe spécial de la classe laborieuse*, a periodical authored by the Parisian labour aristocracy, accidents were not seen as a technical problem, but as the consequence of greed and a symptom of the dysfunctional nature of capitalist society. Its line of argument remained within the terms of the liability regime. After the explosions, the articles demanded stricter safety standards and 'severe' punishments for the bosses. Above all, the authorities ought to include workers in their safety efforts: they, after all, knew the machines best.[51] If the worker was being held responsible, then he should also get to define the production methods.

Some bosses did in fact opt to use their workers' experience to inform workshop safety. Henri Loyer, a textile manufacturer in the north of France, had the following rule posted in his factory: 'As the workers are more used to operating the machines than the masters . . . they are better able to point out the possible causes of accidents. Consequently, I will only consider myself liable to a workman for the consequences of an accident if, after having been warned by him of an imperfection, I refused to carry out the necessary work.'[52] In the mid-nineteenth century, other safety projects were being considered, in addition to those of the Mines administration, involving a different kind of workplace politics.

The hoped-for state of technical perfection always remained beyond reach – and from the 1840s, the crisis of the responsibility principle also

---

50 J. R. Smith, *The Causes of Steamboat Explosions, and the Evils under which They Have Been Permitted to Occur*, 1852.

51 'Ouvriers tués et blessés par l'explosion d'une machine; jugement du maître', *L'Atelier, organe spécial de la classe laborieuse*, 3, 1847, p. 39.

52 Quoted in Alain Cotterau, 'Droit et bon droit. Un droit des ouvriers instauré, puis évincé par le droit du travail (France, XIXe)', *Annales. Histoire, Sciences Sociales*, 6, 2002, p. 1550.

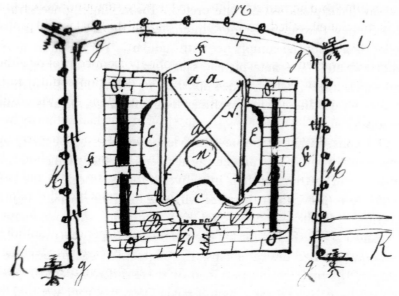

'Parapeur'[53]

spread to the world of engineers. There were many symptoms of this: mounting criticisms of standardisation and the growing complexity of devices, the recognition of new physical phenomena, and statistical evidence of how *normal* accidents were. Reasons were piling up to doubt the technical utopia of perfect machines and responsible humans.

First, the development of knowledge about vaporisation and boiling made it possible to envisage new causes of sudden explosions, and it also increased the uncertainty surrounding boilers. For example, the Swiss physicist Louis Dufour showed that the absence of gas in water produced considerably delayed boiling (up to 170°C). However, the water long left simmering in a boiler could reach such a temperature and then, through some jolt or impact, mix with air and suddenly boil. The amount of steam this abruptly released would cause a devastating explosion.[54]

In the same way, the emergence in the mid-1840s of the concept of material fatigue undermined the project of securing safety through norm-setting. This term, which originated in eighteenth-century metallurgy, was used by engineers to explain the near-invisible fractures in metal components subjected to repeated stress. This research, which initially focused on suspension bridges and railway axles, also changed

53   Archives de la Société d'encouragement pour l'industrie nationale, CME 22, 1831.

54   Dufour, 'Sur l'ébullition de l'eau et sur une cause probable d'explosion des chaudières à vapeur', *Archives des sciences physiques et naturelles*, 21, 1864, p. 201.

the way that explosions were interpreted.[55] The mining engineers trained in these new theories were more interested in the use and wear of the boiler than in formal compliance with standards. For example, an 1866 report explains: 'The metal plate was *fatigued* by the effects of expansion and contraction ... which ended up *irritating* the metal.'[56] The technical vocabulary used here made machines sound like unpredictable animate creatures.

One consequence of this new knowledge was that it gave the worker a more important role in preventing accidents. Rather than controlling workers and restraining their initiatives, the regulations of the 1840s sought to improve their interface with the machine. In 1843, France's Ministry of Public Works ordered the use of a mercury manometer, a float and a glass tube to indicate the water level. In 1847, an order added a second tube and indicator taps. The engineman was encouraged to compare the information provided by these different devices. Redesigning these devices was a sign of recognition that they were fallible. Measuring high pressures and water levels was far from straightforward: foam disrupted the floats, the glass gauges went opaque, the pipes clogged up, and so on. In the 1850s, dozens of models of manometers and level gauges were patented, and manufacturers accused each other of causing explosions. Far from producing responsibility, the deployment of scientific practices of precision in the industrial world blurred the boundaries of blame.

Lastly, the crisis in responsibility was linked to the emergence of a statistical representation of accidents that removed them from a framework of the individual fault. For example, in 1881, as the House of Commons debated compulsory insurance for workers, a lawyer showed that the number of boiler explosions in Great Britain was closely correlated with commercial prosperity.[57] Accidents became commensurable with properly social phenomena.

France's technical safety standards were thus the focus of growing criticism, with the factors of safety imposed by authorities blamed for hobbling manufacturers' own initiatives. According to the formula set by the 1828 norms, large boilers with internal furnaces required extremely thick metal plate. The standard was blamed for slowing down industrialisation: in the 1860s, there were 23,000 boilers in France, compared with

---

55   Stephen Timoshenko, *History of Strength of Materials*, New York: McGraw Hill, 1953, pp. 162–73.

56   *Annales des mines*, 1866, p. 462.

57   Walford, 'The increasing number of deaths from explosions'.

an estimated 90,000 in Britain. So, under pressure from industrialists, a decree in January 1865 'relieved industry of unnecessary obstacles' by abolishing the statutory thickness of boilers. What remained were the legal test, the second valve and the indicators.[58] This decree was decisive, in that it ushered in a new form of risk regulation: the state delegated to manufacturers the task of monitoring and ensuring safety. The Société Industrielle de Mulhouse (which had already criticised administrative standardisation in 1827) immediately founded an Association alsacienne des propriétaires d'appareils à vapeur, modelled on the Manchester Steam Users' Association.[59] Eleven similar associations were created between 1870 and 1900. At the beginning of the twentieth century, some 25 per cent of France's 100,000 boilers came under a system of private inspection.[60] French regulation had caught up with Britain's self-regulated system. This form of self-monitoring was then more generally extended to other kinds of machines: in 1867, the Société Industrielle de Mulhouse set up the first industrial association for the prevention of machine accidents, which was later imitated across France.

5. Railways and *a posteriori* regulation

Railways, because of their complexity, also contributed to casting doubt on the effectiveness of safety standards. Unlike factories, the land on which they were built was simply conceded to the companies and remained in the possession of the state, which was thus in a position to impose numerous constraints. The routes had to be approved by engineers from the Administration des Ponts et Chaussées, and specifications laid down the embankments, crossings, maximum gradients and curves of the tracks.[61]

Yet, the unpredictability of the railway system and the countless circumstances that affected its safety made it impossible to imagine standardising it in a similar manner to boilers or gasometers. There were no parameters or devices deemed critical to safety, in the same way as the

---

58   Rapport à l'empereur sur la fabrication et l'établissement des machines et chaudières à vapeur, Paris, 1865, and Chapuis, 'Risque et sécurité des machines à vapeur au XIXe siècle'.

59   Jean Zuber, 'Sur un projet d'association ayant pour but de prévenir les explosions de chaudières à vapeur', Bulletin de la Société industrielle de Mulhouse, 36, 1866, p. 394.

60   Statistique de l'industrie minérale et des appareils à vapeur en France, Paris: Imprimerie nationale, 1905, p. 103.

61   François Caron, Histoire des chemins de fer en France 1740–1883, Paris: Fayard, 1997, vol. 1; Georges Ribeill, 'Des obsessions de l'État aux vertus des lampistes: aspect de la sécurité ferroviaire au XIXe siècle', Culture technique, 1982, no. 11, pp. 287–97.

thickness of wrought iron plate or the safety valve. So, rather than propose standardisation, the first regulations instead placed the railway companies under a special police force. As early as 1827, the prefect issued a decree for the line between Lyon and Saint-Étienne, the first built in France. At the end of the 1830s, the prefects appointed supervisory commissioners responsible for 'maintaining order' on each new line.

The relations between the police and the railway companies were generally abysmal. Breaches and accidents were reported from one day to the next.[62] Police *commissaires* issued reprimands to mechanics and engineers; they would sometimes forbid convoys from departing on safety grounds or force the company to carry out immediate repairs.[63] Given the capital invested and the financial consequences of delays, the companies refused to be subjected to the orders of a mere police officer who, according to the manager of the Montpellier–Sète line, 'understands nothing, absolutely nothing about the running of an engine'.[64] In 1838, the manager of the Lyon–Saint Étienne line obtained the cancellation of a prefectoral decree. The prefect complained to the minister: 'This goes to show that the company has more power than the *départemental* authorities'.[65] Like the heavy chemicals industry in 1810, rail capital needed freedom from police interference. A specific set of regulations was thus introduced for the railways.

First, the law of 15 July 1845 made the railways subject to exceptional policing rules. Seeing as danger could come from any part of the system at any time, surveillance had to be permanent, total, 'active and almost finicky'.[66] The tracks were completely enclosed and guarded by rail workers. The railways were also liable to severe penalties. On 10 May 1842, the day after the first rail disaster (see below), Charles Dupin called on the National Assembly to impose 'fines and even corporal punishment', since damage payments alone would not be enough to impress rich and greedy companies.[67] The railway police law of July 1845 fulfilled this aim: sabotage was punishable by death, any resistance to a company's agents was

---

62   AD Hérault, 5 S 65, 'Règlement de police pour le service du chemin de fer de Montpellier à Cette', 1839. On the early railways, minor derailments occurred every week: see AD Loire, 5 S 75; 1 M 755; AD Hérault, 5 S 67; AN F14 9566.

63   AD Loire, 5 S 75.

64   AD Hérault, 8 S 195, 15 July 1841.

65   AD Loire, 5 S 75, letter from the prefect of the Loire to the director of Ponts et Chaussées, 18 February 1839.

66   The expression comes from the Minister of Public Works; see APP, DB 492, Chemins de fer, Documents parlementaires, 29 January 1844 session.

67   *Le Journal des débats*, 10 May 1842.

punished as an act of rebellion, and 'blunders, inattentiveness, negligence and failure to comply with the regulations' were punishable by between six months and five years in prison. The Minister of Public Works justified these penalties by citing 'disasters that frighten the imagination'.

Second, an ordinance of November 1846 stipulated regulations for the railways. Unlike the standardisation of boilers and gasometers established by the Académie and imposed on industrialists, the Railways Ordinance was the result of lengthy negotiations between the companies, the Minister of Public Works and a commission made up of engineers from the 'Ponts et chaussées' and Mining administrations. The preparatory work shows that the companies' main concern was to defend their freedom to operate. For example, Article 57 of the draft required the regulations to be posted and distributed to all employees. The companies rejected publicising them in this way because 'basing themselves on knowledge of these regulations, subordinates could wrongly refuse to obey their superiors'. Similarly, one company objected to Article 78, which reiterated the principle of police supervision: 'Without the most severe disadvantages for the regular running of the service, it would not be possible to change the role of public authority agents from passive supervisors to active commanders.'

As far as the engineers were concerned, the prevalent feeling was that the railway system was in its infancy and that they had neither the right nor the competence to set norms for it. For instance, while the draft regulations imposed a uniform system of signals on all companies (to avoid confusion among employees), the commission felt that firms should be left to their own devices so as not to stand in the way of 'the progress to which experience may lead'. The first step was to leave things as they were and then impose what emerged as the 'best system'. Moreover, the engineers feared that the authorities would incur a liability of their own if they laid down too many norms.[68]

As a result, the 1846 ordinance did not define any specific technical forms. For example, Article 2 reads: 'The railway and the works dependent on it shall be constantly in *good condition*,' and Article 12: 'Passenger carriages shall be *of sound construction*.' In fact, rail regulation was based on *a posteriori* control: the companies proposed the layout of the tracks, equipment, operations, timetables, signals, barriers, and so on, and the Ministry of Public Works then gave its approval. The combination of severe penalties and vague regulations ultimately produced a form of

---

68   AN F14 10041.

– Well, the train that's about to arrive just had a narrow escape! The whole convoy would have gone up in flames if I hadn't had my eye on it . . .
– What – what on earth?
– By Jove . . . a pin whose head was right on the rail! Lucky I saw it in time![69]

regulation based on case law. In the wake of accidents, the justice system was responsible for defining what 'good condition' and 'sound' actually meant, and determining which penalties applied.

6. Random disasters

In the 1840s, disasters on the railways transformed the public's understanding of technology. Their repeated occurrence reflected an unexpected aspect of modernity: the mastery of nature through technology turned into a loss of control over technology itself. The public were dismayed to learn that they had to put their lives in the hands of complex systems whose behaviour defied all prediction. A popular book on railways explained that, given the 'complication of the elements of the railway service', disasters 'occur as a result of causes inherent in the system, the effects of which it is impossible for anyone to foresee and avert'.[70] One

---

69   *Le Charivari*, 1843.
70   Arthur Mangin, *Merveilles de l'Industrie, machines à vapeur, bateaux à vapeur, chemins de fer*, Tours: Mame & Cie, 1858, pp. 216–17.

engineer compared the railways 'to an unstable equilibrium that can be upset by even the smallest jolt'.[71] Daumier's caricatures in the *Charivari* humorously illustrated the surge in fatalities resulting from a chaotic technology, in which the slightest causes could produce the greatest catastrophes.

When accidents did happen, the identification of the railways with a chaotic system represented a considerable financial advantage for the companies, which sought to avoid liability by claiming they were the victims of 'act of God' events. Take the debates that followed the first great railway disaster in history. On 8 May 1842, on the Paris–Versailles line, the front axle of the locomotive broke. The boiler spilled flaming coal and the first five carriages tipped over onto the inferno. Trapped in the carriages by the security measures themselves, the passengers were unable to escape. The death toll was counted at fifty-five, with around a hundred injured. The disaster had an enormous impact, dominating newspaper columns for several weeks.[72]

The press called for retribution: the inquest was expected to find the responsible parties, and then the courts would punish them, and the authorities enact regulations to prevent similar disasters in the future. For commentators, the survival of this innovation itself was at stake: 'If the fault can be attributed to men, the cause of the railways, albeit compromised for a moment, will be won';[73] 'if similar mishaps may be repeated, it would be necessary to give up on the use of the railways'.[74] Even for the *Recueil industriel*, regulations had to be issued, which would 'remove the slightest trace of doubt or mistrust . . . Without it, we would have to despair of civilisation'.[75]

In December 1842, the director, chief engineer and station manager were hauled into the dock. From the company's perspective, it was vital that the unpredictable nature of the technology should be recognised, for its capital would not be enough to pay compensation. The determining cause, that is, the broken axle, was not blamed on the company. This was, indeed, a common type of incident on trains in the 1840s.[76] But

71 Félix Tourneux, *Encyclopédie des chemins de fer et des machines à vapeur*, Paris: Renouard, 1844, p. 3.

72 Hélène Stemmelen, 'Une catastrophe technologique au XIXe siècle à travers le journal *Le Temps*', *Culture technique*, September 1983, no. 11, pp. 309–15.

73 *Le Temps*, 10 May 1842.

74 *L'Atelier* no. 9, May 1842.

75 *Recueil industriel*, 1842, vol. 18.

76 Many theories were put forward: magnetic force, oxidation or crystallisation of the metal, or tremors causing microcracks. See *Comptes rendus hebdomadaires des*

this was not enough to plead that this was a merely chance event, as in general such breakages did not produce deadly disasters. Rather, it was the company's job to keep this kind of incident under control. It thus put forward a different definition of chance, based on the *combination* of two misfortunes. Perdonnet, a mining engineer working on the Paris–Versailles train, explained to the Académie des sciences that the axle failure had ended up in disaster because of a 'quite extraordinary meeting' of circumstances. 'For us to have to deplore such a misfortune, there needed to be an almost unheard-of coincidence' whereby 'the axle broke at both ends at the same time, that axle was the front one, the event happened a short distance from a road crossing the track at level, and finally the fire from the furnace spread precisely where the wagons had come together'.[77]

The civil parties rejected this definition of a 'chance event'. If Perdonnet spoke of an extraordinary coincidence, it could easily be explained in terms of the circumstances of the accident. The Versailles fountains show was set for 8 May 1842 – and the rail company expected an impressive attendance. This demanded a longer convoy than usual, and the chief engineer thus placed two locomotives at the head of the train, including an old four-wheeled model. To take advantage of the large number of customers, the manager also encouraged the engineman to step up the pace. The dilapidated state of the locomotive – and especially the absurd idea of locking the passengers into the carriages – turned a minor incident into an utter disaster. To the great surprise of the public, the court accepted the theory of the 'chance event' and sentenced the victims to pay the costs: 'The catastrophe must be attributed to fate . . . whose chances everyone must bear, and to this immense violence, which was insurmountable and impossible to foresee and tame'.[78]

The notion of the 'random disaster' gave risk a new meaning. It became much more than a reflection of human error. The risk integral

---

séances de l'Académie des sciences, 14, 1842, pp. 319, 375, 671, 673, 796, 818. The disaster of 8 May 1842 gave fresh impetus to the study of material fatigue. The interior minister required companies to keep a register of axles, showing the date they entered service and the loads/distances they had carried. See 'Ordonnance provisoire du ministre des Travaux publics, 19 juin 1842', *Journal des chemins de fer*, 1 June 1842, p. 67.

77   *Comptes rendus hebdomadaires des séances de l'Académie des sciences*, 14, 1842, p. 707. Augustin Cournot's definition of chance as the meeting of independent causal chains drew direct inspiration from this judicial debate on the nature of 'fortuitous cause' (*Essai sur nos connaissances et sur les caractères de la critique philosophique*, Paris: Hachette, 1851, p. 52).

78   *Mémoire de la Compagnie du chemin de fer de Paris à Versailles dans le procès relatif à l'accident du 8 mai 1842.*

to technology broke progress out of the mediocre and comfortable materialism within which Romantic writers had sought to confine it.[79] The catastrophe no longer represented an accusation against progress, but instead served to redeem it. It demanded that technical progress be given a higher purpose: the necessary sacrifice that spurred on civilisation itself.

As the company's lawyer Eugène Bethmont made his closing argument, he presented the railway as the field of the brave for the nineteenth century: 'Our fathers died on the battlefields, they died there with glory, they died there for territories ... and we seek our glory and our conquests in industry; we ask it for our glory, our great destinies.' The disaster of 8 May 1842 came at a moment when the parliament was debating a railways act. The situation was extremely embarrassing: the Paris–Versailles company could not be prosecuted without scaring off the Pereires, Rothschilds, Hottingers, Reeds and Laffitte on whom the government counted to finance the emerging rail network. This was clearly not the right time to discuss the liability of railway companies. So the legislature maintained its 'imperturbable calm' – which was itself much criticised.[80] On 11 May 1842, in the Chambre des députés, Lamartine interpreted the disaster in heroic terms:

> We must pay with tears the price that Providence places on its gifts and favours; we must pay with tears, but we must also pay with resignation and courage. Gentlemen, let's face it! Civilisation is also a battlefield where many succumb for the conquest and advancement of all. Let us pity them, let us pity ourselves and let us march onward.[81]

Thanks to risk, progress had become a heroic epic.

7. Bodies insured, bodies paid for

In France, the latter half of the nineteenth century saw the case law on technological accidents turn in a direction much more favourable to victims. After 1842, most rail disasters were blamed on the companies,

---

79 Michael Löwy and Robert Sayre, *Révolte et Mélancolie. Le romantisme à contre-courant de la modernité*, Paris: Payot, 1992.

80 Alexandre Guillemin, *Lamentation sur la catastrophe du 8 mai 1842, au chemin de fer de Versailles*, Paris: Gaume, 1842.

81 Georges Schlemmer and Henri Bonneau, *Recueil de documents relatifs à l'histoire parlementaire des chemins de fer*, Paris: Dunod, 1882.

which were ordered to pay heavy compensation.[82] This case law was based on three principles: (1) any failure to comply with the regulations, however slight, was legally equivalent to fault; (ii) even if a company complied with all the ordinances, it could still be liable if it was in its power to prevent the accident (for instance by employing better equipment or more experienced employees); and (iii) a derailment or collision, when no fault could be identified, in itself constituted sufficient grounds to hold the company guilty.[83] Railway companies soon had to factor the cost of occupational accidents into their financial plans. But the cost was low: in 1842, the manager of the Paris–Rouen railway estimated it at 0.5 per cent of turnover.[84]

A similar jurisprudence was applied to the steam boilers. In 1853, an explosion caused by a faulty supply was blamed on the owner. In 1869, the Cassation Court enshrined this doctrine, basing itself on the liability of the 'custodian of things' as stipulated in the Code Civil: 'An accident caused by an industrial machine, such as a steam engine which has exploded, is presumed to result from the fault of the owner.'[85] Between 1850 and 1880, drawing on the belief that the organisers of production should bear responsibility for accidents, the courts developed a doctrine of the employer's duty to ensure safety. This allowed for compensation to be awarded to the worker, except in cases of gross negligence.[86]

France's first bill on liability for workplace accidents, tabled by the Socialist MP and former bricklayer Martin Nadaud in May 1880, aimed to extend this case law on technical accidents to all trades: 'We propose, gentlemen, to reverse the obligation of proof; today it is the worker's responsibility; in the future we want it to be the employer's responsibility.' As Nadaud pointed out, in reality this was only apparently a legal innovation: since steam and machine technology had reduced the worker to

---

82   Jules Lan, *Les Chemins de fer français devant leurs juges naturels*, Paris: Lacroix, 1867. Employees, including engineers, were often handed prison sentences.

83   Marcel Lacombe, *De la responsabilité des compagnies de chemin de fer en matière d'accidents survenus aux voyageurs*, law thesis, Toulouse, 1908.

84   Bartrip and Burman, *The Wounded Soldiers of Industry*, pp. 70–2. Irish popular science writer Dyonisus Lardner offered a calculation, which he considered reassuring, of the number of workers who met their end per passenger mile travelled; see Dyonisus Lardner, *Railway Economy: A Treatise on the New Art of Transport*, London: Taylor, 1850, pp. 310–20.

85   French parliamentary document no. 1334, 1882, p. 10, and *Recueil général des lois et des arrêts Sirey*, 1871, Court of Cassation ruling of 23 November 1869, p. 10.

86   According to Ernest Tarbouriech, 'it is not necessary for fault to be formally established, it is sufficient for it to be implicit' (*Des assurances contre les accidents du travail*, Paris: Marchal-Billard, 1889, p. 207); Ewald, *Histoire de l'État providence*, pp. 191–220.

'the state of an automaton' or 'an accessory to the machine', they merely demanded the declaration of a new legal presumption.[87]

In February 1882, the moderate republican Félix Faure tabled a counterproposal based on the notion of *occupational risk*. His radical solution was to recognise that most accidents owed not to the fault of either the employer or the worker, but to the 'inevitabilities of the surrounding environment'.[88] In return, the employer was made responsible for no-fault liability and had to take out collective insurance for his workers.

Industrialists rallied behind this new principle; no wonder when it had the considerable merit of reducing their own liability. They feared that a law which reversed the presumptions of proof would confirm, and indeed extend, a body of case law which had already begun to turn against their interests. The railways and steel companies were the strongest supporters of the new theory of occupational risk. Socialists and trade unionists vehemently opposed Faure's plan. According to the programme of Jules Guesde's Parti ouvrier français, since 'the fear of compensation is the beginning and the end of employers' foresight', being able to rely on insurance would make the 'butchering of workers' into an ever more common occurrence.[89]

The 1890s saw the rise of a powerful discourse which promoted the perception of accidents as somehow inherent to technology. This was indeed a 'discourse', because at the same time, statistics from insurance companies and France's mines administration showed that accidents were decreasing considerably; in 1880, explosions caused 3 deaths per 10,000 steam boilers annually, and by 1900 this rate had fallen to around 1 per 10,000.[90] However, the insurance industry promoted its services by stressing the dangers of modern life: progress, that is, the rise of steam, rail and gas, was creating new and inevitable dangers against which the middle classes were unable to protect themselves; only taking out insurance would let them live with peace of mind in a technological society.[91]

---

87 Martin Nadaud, 'Proposition de loi sur la responsabilité des accidents dont les ouvriers sont victimes dans l'exercice de leur travail', parliamentary document no. 2660, 1880, p. 2; Parliamentary document no. 1334, 1882, p. 12.

88 Georges Cornil, *Du louage de services: ou contrat de travail*, Paris: Thorin, 1895, p. 236.

89 The more reformist 'possibilists' (Brousse and Allemane) supported the occupational risk doctrine. See Tarbouriech, *Des assurances contre les accidents du travail*, p. 298; Yvon Le Gall, 'La préparation de la loi de 1898', *Histoire des accidents du travail*, 1981, nos. 10 and 11.

90 *Statistique de l'industrie minérale et des appareils à vapeur en France*, p. 105.

91 See, for example, La Préservatrice's prospectus: *En Wagon. Accidents-Assurance. À propos des grandes catastrophes de chemins de fer*, Paris, 1891.

In 1894, in his inaugural lecture at the École libre des sciences politiques, the engineer and economist Émile Cheysson reiterated the insurers' argument about the *inevitable* dangers of mechanisation.[92]

The international congresses on workplace accidents held regularly from 1889 onwards brought together industrialists, insurers, actuaries and engineers (but no trade unionists). Their aim was to convince legislators of the social benefits of laws based on occupational risk, similar to that which had already been implemented in Germany. Ernest Tarbouriech, a lawyer specialising in insurance law, explained that, given the direction in which liability law was heading, such legislation would in fact be to industrialists' advantage. For the sake of being able to compensate workers, the courts had in fact established an iniquitous jurisprudence: the employer had to prepare for 'not only the usual causes of accidents, but simply the possible ones', 'adopt all the modifications brought about by the progress of industry which result in a reduction of the danger, whatever the expense', anticipating 'the ordinary carelessness of workers, the habit which familiarises them with the danger'.[93] Cheysson confirmed: 'Through a humanitarian fiction the courts endeavour to find fault, to create it even where it does not exist, in order to compensate the victims.'[94]

For their part, employers resented legal proceedings and the prospect of finding themselves in court with a civil party. Above all, they felt that compensation was excessive and unpredictable. In accordance with the judicial principle of individualised and full compensation, the payments varied considerably depending on the circumstances of the accident, the worker's length of service and the number of dependents. This produced considerable financial uncertainties.[95] Worse still, some lawyers encouraged workers to refuse conciliation, pushing them to go to court and taking on all the legal costs in exchange for one-quarter of the compensation. According to the textile bosses' representative, manufacturers were quite prepared to accept a system of lump-sum compensation in

---

92   Janet Horne, *Le Musée social, aux origines de l'État providence*, Paris: Belin, 2004, p. 235.

93   Tarbouriech, *Des assurances contre les accidents du travail*, p. 215.

94   Edouard Gruner, *Congrès international des accidents du travail*, Paris: Librairie de l'École Polytechnique, 1890, vol. 2, p. 237.

95   Tarbouriech gave the example of two boiler explosions. The first, in Marnaval in the Eure *département*, claimed ninety-six victims, including thirty-one deaths, and the company was ordered to pay 180,000 francs' worth of damages, that is, just under 2,000 francs per victim. In Valenciennes, after a similar accident, the Société des Forges had to pay 70,000 francs to compensate for only three deaths. If the court had been as severe in the first case, the firm would have gone bankrupt. See Tarbouriech, *Des assurances contre les accidents du travail*, p. 233.

exchange for an extension of their liability to include chance events.[96] The workers themselves would benefit, since occupational risk guaranteed certain compensation, albeit at lesser levels.

Finally, the insurance companies constantly lobbied for the law. Since the 1860s, they had been trying to sell policies to the working classes, but two obstacles prevented them from doing so: first, since premiums were low, the cost of door-to-door sales was prohibitively high; second, workers' mutual societies (which also served as strike funds) provided them with alternative forms of protection. The insurers were the great promoters and winners of the 1898 law: by making it compulsory for employers to insure workers collectively, it finally opened up the popular security market to them.[97] As compulsory insurance made it commercially pointless to insist on employers' liability, the insurers were now able to present accidents as inevitable and highlight their specific ability to socially distribute the inherent costs of progress.

Contrary to the vision put forward by its promoters – and indeed, François Ewald's interpretation of it – the 1898 law was not simply a law to socialise risk. Rather, behind this social project, industrialists managed to achieve two fundamental political aims. First, the law enshrined a one-sided conception of power in the factory: the boss was implicitly recognised as the sole organiser of production, who in exchange for this took on the industrial risk. This was part of an evolution in labour law that recognised the subordination of the employee to the employer.[98]

Second, the doctrine of occupational risk was the culmination of Chaptal's old project of providing a stable footing for the industrialist's activity. The essential function of occupational risk was to make the cost of accidents calculable. At the International Congress on Industrial Accidents in 1897, an engineer explained that the advantage of occupational risk was that it eliminated unforeseeable legal costs: in his words, it was better to have 'provisions that may be less equitable, but

---

96   See C. Jouanny, 'État actuel de la question des accidents du travail dans les syndicats professionnels en France', *Congrès international des accidents du travail et des assurances sociales*, Brussels: Weissenbruch, 1897, p. 584. Jouanny, as the textile industry's delegate, represented 15,000 employers. He was also a member of the Committee of the Conference on Workplace Accidents.

97   Daniel Defert, 'Popular life and insurantial technology', in Burchell and Gordon (eds), *The Foucault Effect: Studies in Governmentality*, London: Harvester, 1991, pp. 211–33.

98   This law established the transition from *louage d'ouvrage* – a system based on providing a specific product, which did not provide for any subordination of employee to employer – to the temporal leasing of the worker's services. Cottereau, 'Droit et bon droit', and Cornil, *Du louage de services*.

are undoubtedly simpler and more practical'.[99] In his opening speech, the president of the French insurers' union presented a calculation of the economic cost of the law that the National Assembly was about to vote on. Using actuarial tables and the scale of payouts, he arrived at a figure of exactly 82,978,416 francs a year, a rather trivial sum compared to French industrial production. The value of the law could thus be assessed not only in legal or ethical terms, but also in terms of national economic competitiveness.[100]

It would, then, be a mistake to see occupational risk as a 'social' law imposing regulation on the previously laissez-faire world. Rather, it was the solution promoted by industry and insurers to the crisis of a previous, much more restrictive form of regulation, based on safety standards, the apportionment of blame and compensation for damage. The project of technical perfection through standards had in the 1860s led to constraints which industrialists considered incompatible with the competitiveness of national industry. Yet, in order to do justice, the courts had developed a doctrine of responsibility based on a perfect boss, who was as much a fantasy as the perfect machines of the mining engineers' imagination.[101]

Conversely, by recognising that accidents were intrinsic to technological society, the doctrine of risk made it possible both to liberalise technical forms and to govern them more effectively. The economy could now be thought of as a singular whole, which could be optimally managed by setting lump-sum compensation payments. The consequence of this overarching power, which no longer bothered to do justice to individual accidents, was to subject the worker's body to a new and meticulous accounting mechanism. Scales, drawn up jointly by insurers, judges and industrialists, defined the value of each piece of the worker's body according to his occupation.[102]

In 1881, Félix Faure's proposal had caused a scandal in the National Assembly because it seemed to reduce the worker to a mere factor of production. One deputy rejected a law which, in his view, amounted to

99   Gruner, Congrès international des accidents du travail, p. 68.

100   Ernest Tarbouriech, La Responsabilité des accidents dont les ouvriers sont victimes dans leur travail, Paris: Giard, 1896, p. 115.

101   In the world of steam engines, civil engineers and contractors complained even about the few administrative provisions that did still remain in force after 1865. They wanted the government to abolish all interference with standards, in exchange for them taking on full responsibility. See François Hervier, Les Explosions de chaudières, examen critique des moyens préventifs, Paris, 1894, p. 21.

102   A. Duchauffour, Les Accidents du travail, manuel de conciliation, Paris: Baillière, 1905.

'treating the worker as a mere thing'. The mining engineers also opposed the insurance solution, which they considered a despicable abandonment of responsibility at the expense of workers' safety. But this solution did indeed prevail. A powerful legal, economic, insurance and 'reforming' discourse on accidents had succeeded in making morally acceptable the normalisation of accidents and the integration of the worker's body into the company's economic calculations. Félix Faure offered this striking analogy: 'Everyone knows what is meant by the overheads of an industrial company. They include rent, fire insurance, equipment repair and maintenance costs and a sum intended to represent their amortisation. Well, add to that the repair and amortisation of *human equipment*.'[103] At the end of the nineteenth century, the industrial world – one of metal and movement, of death and mutilation – was paradoxically judged neither good nor bad; the violence behind it was so to speak, neutralised. The liberal principle of fault existed in no more than residual form: the functionality of the capitalist system had itself rendered it obsolete. Just like it had done to the 'surrounding environment' previously, capitalism now succeeded in encompassing workers' bodies within its logic of financial compensation. Contrary to Ewald's thesis, occupational risk did not point towards the birth of the 'welfare state', but towards a new form of laissez-faire, one all the more effective because it dispensed with morality.

According to Walter Benjamin, the concept of progress has to be based on the idea of catastrophe: 'The catastrophe is that things continue as before.'[104] Disasters were indeed the matrix of the discourse of progress, which became culturally dominant in the second half of the nineteenth century precisely to compensate for the loss of the ideal of technical perfection and to 'carry on as before' despite the disasters that had repeatedly occurred. The discourse of progress, which exalted humanity's grandiose goals, also served to exorcise new and immense anxieties. The promises of the future justified the random victims lost along the way. In this respect, the discourse of progress is the cultural expression of capitalism's turn towards insurance.

Jules Verne's novels offer a good illustration of the evolution of discourse on technology and reflect the success of the stochastic, insurance-based interpretation of accidents from the 1870s onwards. In *Five Weeks in a*

---

103   Quoted in Tarbouriech, *La Responsabilité des accidents dont les ouvriers sont victimes dans leur travail*, p. 111.

104   Walter Benjamin, *Charles Baudelaire: A Lyric Poet in the Era of High Capitalism*, Paris: Payot, 1982, p. 342.

Balloon (1863), the hot-air balloon had been described as a perfect object, meaning that all blame for any accidents was to be apportioned to the pilots. It belonged to a deterministic universe: 'Everything that happens in this world is natural; now, anything can happen, so we must be able to foresee everything.' Safety was based on physical laws: Verne gives the weights of the characters and objects on board; with this table and the Archimedean principle, his balloon could fly over Africa with complete peace of mind, 'there was not a single objection to make; everything was foreseen and resolved'.[105] The paradox of Verne's work, with its parade of fabulous machines and their ultimate destruction, becomes clearer if we consider it within the discursive regime of the responsible man: the novelist imagines perfect techniques in order to create tragic heroes, masters of their own destinies, like Nemo sinking with the Nautilus that he has launched against a battleship.[106]

In contrast, *Around the World in Eighty Days*, published ten years later, is a novel about probabilities. Phileas Fogg is a 'statistical' man: 'He often corrected, with a few clear words, the thousand conjectures advanced by members of the club . . . pointing out the true probabilities, and seeming as if gifted with a sort of second sight.'[107] The theme of probability is omnipresent throughout the novel, with the card game whist, in which Phileas Fogg excels, and the wager, which is the theme of the entire adventure. In making his bet, Fogg was not counting on the perfection of modern means of transport: '"Yes, in eighty days!" exclaimed Stuart [. . .] "But that doesn't take into account bad weather, contrary winds, shipwrecks, railway accidents, and so on." "All included," returned Phileas Fogg, continuing to play despite the discussion.'[108] In Verne's later novels, accidents did not point to fault or hubris, but became an integral part of technology.[109] Risk had redemptive properties: it imbued progress with heroism, turned technology into an adventure, and freed it from any purely materialistic character.

But, if risk was now accepted, it was still necessary to show that it was worth taking. Hence the literary annexation of Africa and Asia: the dangers of these pre-technical and barbaric places served as a

---

105   Jules Verne, *Cinq Semaines en ballon*, Paris: Hetzel, 1865, p. 69.

106   Jacques Noiray, *Le Romancier et la Machine, l'image de la machine dans le roman français (1850–1900), Jules Verne, Villiers de l'Isle-Adam*, Paris: José Corti, 1982.

107   Jules Verne, *Around the World in Eighty Days*, London: Samson Low, Marston, Searle, & Rivington, 1876, p. 3.

108   Ibid., p. 18.

109   *Les Cinq Cents Millions de la Bégum* (1879) ends with the – fortuitous – explosion of a shell due to 'a mysterious molecular action'.

The safe bubble of technology in a barbaric and dangerous world[110]

counterpoint to those of industrial civilisation. Technological isolation was an essential theme in *Voyages extraordinaires*. *Five Weeks in a Balloon* and *Around the World* recount the adventures of Europeans protected from the dangers of barbarism by safe technological bubbles. In *Around the World*, it was the train that allowed Europeans 'in white sheets' to cross the India of wild beasts and the America of buffaloes and Sioux. Similarly, the perfect balloon in *Five Weeks* allowed them to soar above the dangers of Africa.

Verne's novels were a simple reflection of a new representation of the world which now dominated in the West. Geography, hygiene and colonial medicine, as well as adventure and exploration novels, all helped to produce the image of a safe, technological Europe amid a barbaric and dangerous world, thereby justifying the risks of modernity.

In technical and medical journals, articles on vaccination, derailments, explosions and pollution ran alongside accounts of the dangers of Africa and Asia. Readers of the *Annales d'hygiène* would find, in the same pages, articles on colonial medicine, on the mortality of Eastern populations, on the appalling diseases discovered in Africa, statistics on the health of troops in Algeria and mortality in Paris, and reports on the insalubrity

---

110   Illustrations by Édouard Riou and Henri de Montaut (*Cinq Semaines en ballon*), Alphonse de Neuville and Léon Benett (*Le Tour du monde en quatre-vingts jours*).

of certain factories. The risk set Oriental, European, urban and industrial climates in the same statistical universe, and demonstrated that mortality was falling at the same rate that civilisation advanced. This global risk accounting provided the most general possible justification for industrial civilisation.

Conclusion

Liberal capitalism is marked by a curious contradiction. On the one hand, it proclaims that it is impossible to change society. On the other, it happily goes on with a radical transformation of nature. It combines a supposedly realistic acceptance of human goals and social organisation such as they are, with a deeply utopian project of mastering and transforming the world. The cosmos of capitalism consists of individuals whose drive to pursue wealth is deemed unchangeable, and who are situated within a world of infinite technological adaptation. When George Bush Sr declared at the 1992 Rio Earth Summit that 'the American way of life is non-negotiable', he implied that nature and its preservation were negotiable. In conclusion, I would like to show how this destructive reversal has prevailed since the end of the eighteenth century.

The modern disinhibitions which we have studied in this book have taken two forms: one based on moral attitudes, the other on descriptions of the world. The first has sought to produce technophile subjects ready to defy their scruples and their uncertainties in the name of creating a modern, comfortable, healthy and reasonable society. The risk involved in the inoculation controversy is emblematic of how individuals were morally refitted in the manner which technological action required. In the mid-nineteenth century, the rhetoric of progress also had the function of inspiring the courage that modern life required. To counter the traumatic effect of disasters, it anchored in European culture a sense of its superior technology and security, doing so through a comparison between two worlds: the world of before (or elsewhere); and the modern, technical, Western one.

But technophile excitement over progress, orientalism, the promise of a rosy future, or the calculation of probabilities, were probably not decisive in setting technology in motion. The inoculation that was supposed to stake its claim by force of reason – by demonstrating its probabilistic advantage – failed miserably. As for the idea of progress, already in 1932, Lewis Mumford wrote that it was 'the deadest of dead ideas'. The fact that, from Baudelaire to the postmodernists, the discourse of progress has so constantly been on its way out, indicates that it was probably not a strategically decisive element in the modernist project. In the 1970s–1990s, in a period of high technophilia, philosophers and sociologists agreed in again pronouncing it dead (which they welcomed), and this in no way prevented the accelerated artificialisation of the world. The fact that the term *progress* has lost its sheen simply reflects the general acceptance of its logic: today's so-called knowledge-based societies, are obsessed with innovation and technical mastery. Progress has lost its political meaning only for want of real enemies.

Yet what was absolutely fundamental in the nineteenth century – and undoubtedly remains so today – is the ontological form of modern disinhibition. By this, I mean all the descriptions that aim to adjust the world to the technological development imperative. Indeed, it soon became clear that modernisation would run much more smoothly if, instead of relying on the courage of individuals, their fears were simply circumvented by producing anxiolytic description of the world transformed by technology. The modernisers' essential task was not to create individuals who wanted machines, but rather to make machines desirable. Disinhibition would be all the more effective if it crept into the immediacy of our relationship with the world and with technology, through little mechanisms able to cover in silence the violence and the political significance of the objects in question. The aim was to cushion technological shock and neutralise the critical sense of the accidents that occurred; to capture, direct and align perceptions and behaviour, in the direction of technology.

This form of disinhibition through ontological means was dominant because, unlike the sacrificial ethic of progress, it responded perfectly to the political demands of liberalism. For liberal philosophers, the virtues and goals of the individual are reduced to a meagre affair: self-preservation in Locke, the 'calm desire for wealth' in Hutcheson, or the 'security of private pleasures' in Constant. Kant, in his *Project for Perpetual Peace*, explained that a good constitution could even pacify 'a people of demons'. Liberalism vouches for the immutability of human goals of enrichment: if a government intended to base the law on the virtue of individuals, or worse, wanted to produce better, more courageous, more altruistic, more thrifty individuals, it would be exposing itself to serious disappointment. But this conception of politics, which gives up on governing individual goals, came up against the limits of material abundance. The only conceivable solution was thus to increase production and wealth: social harmony would have to be achieved through material progress.

Hence the fundamental role of technology and economic development. As the success of liberal government depended on material prosperity, technology became a *raison d'état*. Around 1800, for those who governed a population, it was imperative to make the interventions that would maximise its biological and material productivity. The stories in this book shows that power has invariably played a decisive role in initiating large technological projects: naked power over orphans transformed into test subjects and vaccine reservoirs; power over peasants who challenged the environmental disorder caused by industrialisation; power that imposed technological risk on city residents. An innovation of any importance will confound society with the new problems and unforeseen situations which it produces: inoculation that fosters epidemics; the chemical industry that destroys crops; and steam engines or railways that kill at random. In 1690, Antoine Furetière defined innovation as 'changing a custom, something that has been established for a long time. In good politics, all innovations are dangerous'.[1] The innovations studied in this book could not be easily subsumed under previous legal rules. They defined states of exception, they suspended the traditional norms governing health, property, environments and the attribution of damage and responsibility.

The nineteenth century thus experienced a constant state of technological exception. Every time a major innovation appeared, it was not integrated into prior regulatory forms, and even less so into forums for public discussion that would pacify social relations and resolve oppositions. Innovation is political not in the sense that it produces externalities, mobilises different stakeholders, or prompts public discussion, but because it polarises the social: it produces winners and losers, profits and ruined harvests, experimental subjects and vaccinees, satisfied consumers and mutilated workers. Contrarily to the utopian views of a so-called technological democracy, the benefits and risks of technology were never debated in the abstract; there was never a moment of dialogue when individuals clashed in the ethereal sky of argument.

---

1   Antoine Furetière, *Dictionnaire universel*, The Hague, 1690, entry 'Innovation'.

The fundamental historical point is that technology has shaped the modes of its regulation, far more than the other way around. Discussion of technology exists, but it only takes place *after* the first complaints or accidents, and therefore after technology has become a *fait accompli* and is already subsumed under *raison d'état*. Its regulation therefore takes place at a time when the usual normative practices are suspended, and generally results in the normalisation of a state of exception. What did the decree of 1810 do if not ratify the environmental exception represented by the chemical industry? What does technical standardisation do, if it does not legitimise objects that were initially perceived as unacceptable?

But, at the same time as the technological imperative, the utopia of soft power was emerging. The importance of this theme in the political thought of the late eighteenth century can hardly be overestimated. A good ruler is one who does not push his people around. By giving free rein to the inclinations of his subjects, he increases the population, the wealth of his kingdom and therefore his power. His gentle approach does not promote insubordination, for it produces gentle subjects: according to Montesquieu, 'eight days in prison, or a small fine, affects the mind of a European brought up under a mild government, as the loss of an arm intimidates an Asiatic'.[2] His gentleness produces subjects who are more sensitive and thus more easily governed.

Hence power's new aim: not to act by coercing bodies, but to guide its subjects' minds through the manifestation of reason. As d'Holbach so aptly put it, governing with gentleness consisted of 'turning weak minds to the reason of which they are ignorant' and thus commanding 'reasonable, docile and truly attached subjects'.[3] Condorcet also emphasised the importance of the sciences, and more particularly statistics, for the post-revolutionary order: 'When a revolution ends . . . one needs to *chain men to reason* through the precision of ideas and the rigour of evidence'.[4] The gentleness of power thus has its correlate in its investment in the field of reason, proof and truth.

Legal scholars have often reflected on the legal nature of the state of exception and the possibilities of escaping from it. If we think of

---

2   Montesquieu, *Persian Letters*, letter 80. According to this same author, 'the gentleness of government contributes marvellously to the propagation of the species' (ibid., letter 122).

3   D'Holbach, Éthocratie ou le gouvernement fondé sur la morale, 'Avertissement'.

4   Marie Jean Antoine de Condorcet, 'Tableau général de la science qui a pour objet l'application du calcul aux sciences politiques et morales', *Journal d'instruction sociale*, vol. 1, no. 4, 1793, p. 109.

innovation as a state of exception, it is generally concluded by prop-
ositional means: the technological state of exception ends through a
redefinition of the fabric of the world, the beings that make it up and
its regularities, as well as a redefinition of the technical forms. After the
technological *coup de force*, science is called on to continue its work of
describing beings, in order to assimilate the exception, to maintain a
society in which technology was *neutralised*, and to restore the liberal
system to its fullness and coherence. The sovereign allied with science thus
also assumes an ontological function. His real power is that of imposing
meanings capable of reducing the divide which innovation opens up.
'Caesar is lord also of grammar.'[5]

This is the root of the new political usefulness of scientists. From the
end of the seventeenth century, when natural philosophers invented
experimentation as a mode of apprehension fundamentally superior to
common experience, scientists have acquired a socially recognised ability
to define the beings that make up the world and their interrelations. The
ontological power of science was thus deployed to found a new form of
political legitimacy that depended on the invocation of the natural order.
Doubtless one of the most famous examples of this is physiocracy, which
sought to establish political economy as an extension of the economy of
nature. This form of factual power manifested itself each time that scholarly expertise purported to reform social practices through a redefinition
of nature. For example, in 1770, when Fougeroux and Tillet studied the
botany of kelp and, on the basis of their report, the monarchy abrogated
the rules that had governed its use for several centuries, this constituted
a political application of the new power conferred by the capacity to
state facts.

This mode of government by scientists is our second invariant. Botany
legitimised the intensive use of nature. Clinical science defined vaccines
in such a way as to ward off any reticence. Hygienism played down the
medical importance of the environment in order to cast industrialisation
as harmless. Then there were the safety standards that blamed the violence
of the industrial world on mistakes by workers. The so-called laissez-faire
of the nineteenth century was only possible because of a very deliberate
attempt to define nature and the forms of technology.

---

5   Carl Schmitt, 'Les formes de l'impérialisme dans le droit international' (1932),
Du politique. Légalité, légitimité et autres essais, Puiseaux: Pardès, 1990, p. 99 (English
translation, 'Forms of Modern Imperialism in International Law', in *Spatiality, Sovereignty,
and Carl Schmitt*, ed. Stephen Legg, London: Routledge, 2011.

According to Saint-Simon, the post-revolutionary power needed to mutate from a 'government of men' to an 'administration of things'. To paraphrase Saint-Simon, we could say that the new liberal power was based *on the government of men through the administration of things*. The fundamental connivance that links science and technology to the liberal order lies in the notion that by defining things and nature in a suitable way, or by amending technology and its uses, we can do without politics.

With the benefit of historical hindsight, we see these mechanisms as crude ruses (such as the field of pustular diseases, in the case of vaccinia). But, at the time, they dominated the apprehension of the world. They could do so because the fog of statements, ontologies and defence mechanisms, which were used to protect innovation from criticism, was established and kept in place by powerful institutions. The science that enabled the liberal exercise of power was set up in this function by authoritarian means. For example, the vaccinators were able to maintain the philanthropic definition of the vaccine only because the state had made its power of censorship available to them, along with the bodies of foundlings as an experimental stage. Knowledge was only capable of transforming the world to the extent that it was enlisted by already established powers.

After 1800, given the overriding importance of technology to the state, the existing judicial, procedural and dialogical forms of truth production (the medical faculty assembly, the public sphere, consultations of trade bodies) were replaced by scholarly and administrative institutions. The trades were no longer sources of knowledge that guided the government but were themselves to be reformed under the orientation of the learned institutions. The public was no longer a body to be convinced nor one which should pass judgement, but a mass to be conducted towards the correct decision. For the government to intervene on its population and its economy (and how could it not intervene, when science and technology offered so many ways of increasing its power?), it had to circumvent the necessarily polyphonic public space, bypass the variety of values and knowledges and, to this end, establish administrative and scientific bodies that would produce the right decisions. Technological committees (such as the vaccine, salubrity or steam engine committees) had a mandate to establish the cognitive and normative frameworks necessary for biopolitics and industrialisation. Technology had acquired too much importance for the definition of its effects to remain a subject of debate.

Society was also seen in a profoundly different light: not as a set of constituted bodies with varying interests and knowledges, but as a sum of individuals whose confrontations had to be regulated. It was no longer a matter of deciding between divergent interests and knowledges, but of creating the right framework for conflicts: their proper level, their means of proceeding, their objects, and above all the cognitive order in which they took place. For example, by redefining the changes to *circumfusa* as a mere inconvenience, the hygienist administration left it up to the courts to arbitrate minor disagreements between individuals. The real oppositions had already been resolved in advance by the scientific redefinition of the links between environment and health. In the same way, because the boiler that exploded had been standardised and perfected by a learned administration, there was bound to be someone to blame and claim compensation from. Hygiene or safety standards thus made it possible to resolve the conflicts of modernity simply as individual disputes. The perfect vaccine defined by philanthropists, the gasometer standardised by the Academy of Science, the steam engine managed by the Mines Administration and the transformation of the environment rendered insignificant by hygienist theories – these all belong to the same historical configuration, which, from the 1800s onwards, claimed to transform the way we live both without risk and without any question of politics.

Historians have shown how liberal political philosophy was ultimately an anthropological project aimed at creating an egotistical, calculating subject, as against the traditional values of gift, sacrifice and honour.[6] We might add they have failed to see that in return the *homo economicus* demanded a world tailored to his own standards – rethought, reconstructed and redefined so that the quest for the greatest utility could be freely exercised. At the beginning of the nineteenth century, science and technology adjusted ontologies and objects with the aim of creating an 'economic world': a *mundus economicus*.

In this book, I did not make the absurd pretence of casting the societies of the Industrial Revolution as respectful of ecological concerns. The aim was not to say that we have always been in a *risk* society or a *reflexive* society. Of course, the societies that set in motion the carbonisation of our economy and atmosphere and caused massive suffering to workers were *not* reflexive. But it is also doubtful whether contemporary societies really are. One of the challenges of this book is to point out certain weaknesses in the sociology of risk and to replace its major diachronic

---

6   Albert O. Hirschman, *The Passions and the Interests: Political Arguments for Capitalism before Its Triumph*, Princeton, NJ: Princeton University Press, 1977; Christian Laval, *L'Homme économique. Essai sur les racines du néolibéralisme*, Paris: Gallimard, 2007.

oppositions with a historical narrative, thus cancelling out the comforting illusion that we are exceptional. If this history were to be extended, it could be presented as that of the successive transformations of modern modes of disinhibition.

After the end of the nineteenth century and the creation of an insurance-based capitalism protecting technological modernity from the legal threats of fault and liability, the 1920s to 1950s were a fundamental stage, marked by the invention in the United States of what has come to be known as the consumer society.

Historians have clearly shown the programmed, strategic nature of this transformation of capitalism. The explicit common goal of politicians, economists, industrialists and advertisers alike was to create a market capable of absorbing the new productive capacities of the Taylorist factories. To achieve this, values had to be transformed: repair, thrift and saving were presented as obsolete habits harmful to the national economy, while repeated and conspicuous consumption, fashion and product obsolescence became respectable objectives. A revolution in advertising (from product advertising to the glorification of consumption as a way of life and a marker of social normality), wage increases and, above all, the introduction of consumer credit were the pillars of a renewed social control. In exchange for consumption, the individual had to accept an increased routinisation of his work and his dependence on credit.

The term *consumer society* thus refers to a new relationship with objects and the environment, and a new form of social control that makes this relationship desirable. Disciplinary hedonism played (and continues to play) a fundamental role in the acceptance of mass production and its disastrous environmental consequences.[7]

At the same time, several phenomena have made these harmful effects socially invisible. Firstly, thanks to the development of oil and transport, the extraction of mineral and energy resources – and, in fact, all the most harmful production processes – have been able to be relocated to far-away areas. As a direct consequence, the suburbanisation movement allowed for a greater separation between residential areas and production sites. For a society undergoing a tertiarisation process, industrial production and its environmental effects took on a much more abstract character.

---

7 Stuart Ewen, *Captains of Consciousness: Advertising and the Social Roots of Consumer Culture*, New York: McGraw Hill, 1976; Lendol Calder, *Financing the American Dream: A Cultural History of Consumer Credit*, Princeton, NJ: Princeton University Press, 1999; Giles Slade, *Made to Break: Technology and Obsolescence in America*, Cambridge, MA: Harvard University Press, 2006.

The intellectual world also gradually lost interest in the material conditions of production. The case of economic discipline, which became the dominant mode of training the social elite, is exemplary in this regard. Marginalist theories turned away from studying the factors of production (labour, capital and land) and focused on the subjective conditions of consumers and producers seeking to maximise their individual utility.[8] This gave rise to a new object of thought: *the economy*, understood as the whole set of relationships defined as economically relevant, and bearing a tenuous relationship with natural constraints.

The idea of *growth* was itself a result of this transformation. Before the 1930s, this notion was bound to a material process of expansion: increasing the production of some material, or by opening up new resources or new territories to the economy. With the crisis of overproduction in the 1930s, growth was rethought, not in material terms but as the intensification of the web of monetary relations. The abolition of the gold standard in the 1930s – that is, the end of the idea that banknotes represent gold – and the invention of GDP by national accounting, completed the dematerialisation of economic thinking. John Maynard Keynes described this turn perfectly. Whereas, in the second half of the nineteenth century, many economists (Stanley Jevons among them) had worried about the depletion of coal reserves, Keynes explained that running out of coal would have no consequences: to ensure employment and economic prosperity, all the British Treasury had to do was to bury banknotes and ask miners to go and fetch them! Thanks to its dematerialisation, the economy could finally be conceived of as growing indefinitely, outside natural determinisms and without altering physical limits, all thanks to its proper oversight by expert economists.[9]

Alongside these indirect modes of government, which succeeded in enlisting the whole of society in a capitalist growth project, more traditional and self-perpetuating forms of the imposition of technology by the state persisted. The French hydroelectric programme that led to the submergence of numerous Alpine villages after the Second World War is a good example of this. Similarly, in 1974, the French electro-nuclear programme was forced through in a unilateral fashion. EDF and the CEA were determined to build as many plants as quickly as possible, to make

---

8   Daniel Breslau, 'Economics Invents the Economy: Mathematics, Statistics, and Models in the Work of Irving Fisher and Wesley Mitchell', *Theory and Society*, vol. 32, no. 3, 2003, pp. 379–411.

9   Timothy Mitchell, 'Fixing the Economy', *Cultural Studies*, vol. 12, no. 1, 1998, pp. 82–101.

the choice in favour of nuclear power irreversible. All means at hand were put to this end: bypassing administrative authorisation procedures, police violence against protesters and massive financial compensation for the local authorities who accepted the reactors.

In the 1980s, this figure of the modernising state, which knew best what was the common good and could thus impose it over the heads of locals, was politically weakened by criticism from both neoliberals and advocates of a technological democracy. The heart of the modernisation process then shifted to the market. The history of GMOs in the United States is thus that of business imposing its will. Against a backdrop of heightened economic competition with Japan and in the wake of the oil crisis, biotechnology seemed to represent a major economic opportunity. In 1980, a Supreme Court ruling overturned all previous case law and accepted the patentability of living organisms. Capital poured in. Genetic engineering became the model for innovation in a neoliberal society: research driven by the private sector, backed by venture capital and producing immediate profits.

Ideologically, biotechnologies marked an important turning point. For they certified the victory of man's unilateral relationship with other natural beings, indeed with life itself, which had to be optimised in the manner of a simple production technique. Postmodernist discourses on transcending the limits between nature and culture, and the aesthetics of hybridisation characteristic of 1980s–1990s philosophy (as proposed by Donna Haraway or Bruno Latour, among others) only confirmed a deepening modernist project of the mastery of nature. For behind the symmetry between humans and non-humans, it is once again man who finds himself transforming nature for his own benefit.[10]

Far from having become reflexive, our contemporary societies fetishise innovation as never before. They have made it synonymous with growth, and political parties of both Left and Right are turning it into a national agenda. Since the 1980s, all financial regulations have been transformed to make economies more flexible, more competitive and more innovative. The growing importance of the private sector in producing innovation, the subordination of scientific research to the demands of economic profitability, and companies' need to come up constantly with new products are increasing the powers of capitalism in defining our technical destiny, to the detriment of democratic control mediated by the state and public

---

10   Hervé Kempf, *La Guerre secrète des OGM*, Paris: Seuil, 2003; Christophe Bonneuil and Frédéric Thomas, *Gènes, Pouvoirs et profits*, Versailles: Quae, 2009; Jean Foyer, *Il était une fois la biorévolution*, Paris: PUF, 2009.

research. More than ever before, science has itself become a business, driven by financial priorities at odds with the precautionary principle. The economic success of biotech firms and the spread of nanoproducts demonstrate, if proof were needed, the intrinsic link between financial profitability, via Nasdaq and venture capital, and the modernist project of artificialising the world.[11]

The second fundamental development at the end of the last century – economic globalisation – has enabled rich countries to offshore some of the risks of industrial production. Developed societies are not running scared of technology; they have simply managed to externalise its most negative consequences, to outside the West. Since multinationals have been relocating not only industrial production but also research and development, in search of lowered wage costs, neither progress nor its control are now the prerogative of the long-industrialised countries. Globalisation makes almost touchingly naive the theory of reflexivity formulated by philosophers and sociologists from now-marginalised Europe.

Finally, this double process has been accompanied by a whole range of regulatory instruments, ideologies and illusions. I will give just a few examples of this.

The effects of neoliberalism on environmental law are well known: the Reagan presidency saw the dismantling of the progress of the previous decade, and the Bush presidency eased the conditions for oil and gas exploitation. Generally speaking, we have moved from a regulatory approach in the 1970s (bans, fines) to a governance approach, guided by economics and market instruments (pollution rights and self-regulation by companies).[12]

Let us take the notion of thresholds with regard to carcinogenic substances. At the end of the 1940s, toxicologists warned governments that, at no matter what dose, certain molecules produced by synthetic chemistry would increase the risk of cancer. A consensus formed that such molecules should be removed from the food supply. In the United States in 1958, the Delaney clause banned the presence of pesticide residues in food. But ultimately the principle that won out in regulatory bodies

---

11 On the evolution of the technosciences in relation to neoliberal economic logic, see the sharp analyses by Dominique Pestre: *Science, argent et politique. Un essai d'interprétation*, Versailles: INRA éditions, 2003, pp. 77–118 and 'Des sciences et des productions techniques depuis trente ans. Chronique d'une mutation', *Le Débat*, 2010, no. 160, p. 115 and more broadly: David Harvey, *A Brief History of Neoliberalism*, Oxford: Oxford University Press, 2005.

12 Neil Cunningham, 'Environment law, regulations and governance: shifting architectures', *Journal of Environmental Law*, vol. 21, no. 2, 2009, pp. 179–212.

in the 1970s was cost/benefit analysis (a risk is tolerated depending on the economic interest of the substances) and the definition of thresholds. New international standards, such as *acceptable daily intakes* (ADIs) in the case of food and *maximum permissible concentrations* (MPCs) in the case of air quality, operated a subtle disguise. Given the non-existence of a threshold effect, these standards in fact enshrined the toleration, for economic reasons, of an acceptable cancer rate.[13]

The term *sustainable* plays a similar role in the increasingly intensive exploitation of nature. The history of fisheries is exemplary in this regard. The principle of maximum sustainable yields, introduced in international treaties after the Second World War (at the 1955 Food and Agriculture Organization conference), enshrined the principle that it was possible, without worry, to fish in optimal quantities that would preserve this resource. Fairly simple ecological models supported a radical increase in the volume of fish caught, from 20 million tonnes in 1950 to 80 million in 1970. But the models defining the sustainable use of stocks did not consider certain factors such as population structure or the degradation of marine ecosystems – and in the space of a few decades they led to the widespread collapse of fish stocks.[14]

Recently, the notion of sustainability has been transformed into a powerful anxiolytic for conscientious consumers. Companies have been quick to grasp the benefits of this malleable category and of environmental certification, since it is always possible to find or create a label that vouches for the sustainability of their production practices.[15] Despite its crudeness, this disinhibition of consumerism has rapidly conquered market shares. The main problem with the notion of sustainability is that it produces the illusion of an effective reconciliation of environmental imperatives and economic efficiency, of control of growth, and of nature being placed under the good governance of companies and certification agencies.

With the rise of the climate question, the whole Earth has been subjected to this same principle of optimising nature. Economists have rethought the climate as an atmospheric resource whose net present value

---

13   Soraya Boudia and Nathalie Jas, *Powerless Science? The Making of the Toxic World in the Twentieth Century*, Oxford: Berghahn Books, 2014.

14   Philippe Cury and Yves Miserey, *Une Mer sans poissons*, Paris: Calmann- Lévy, 2008.

15   An extreme example: wood from plantations created after the destruction of primary forests by napalm in Tasmania was awarded an eco label. See amisdelaterre.org/ IMG/pdf/Certifying_the_Incredible.pdf. See also 'Mauvais génie de la forêt', *Le Monde*, 8 April 2011, on the role of consultants McKinsey in evaluating REDD projects.

they can maximise by defining optimal CO_2 emission curves. Global change is thus translated into a problem of maximising economic growth under climatic constraints. Established in 2007, carbon credits have collapsed and then risen again; they will no doubt continue to fluctuate and circulate, without sufficient questioning of their material reference points, partly because the environmental audit firms that estimate the CO_2 emission reductions by so-called clean development projects have no interest in judging them too harshly. But no matter: their mere existence and exchange are enough to confect the prospect of an economy finally made ecological.[16] It may be feared that these techniques for optimising nature are no more than a decoy that comes instead of a truly controlled human presence in nature.

The 1990s sociology of risk deserves credit for naturalising the idea of public participation in technoscientific choices. The emphasis it placed on uncertainty and the importance of opening up decision-making procedures to a wide variety of stakeholders, and the ideal of a strong democracy and its application in the field of technological choices, have made it possible to make natural an idea that seemed strange thirty years ago. That is, the understanding that the public should have its say, that there are situated knowledges, lay knowledges and useful knowledges drawn from experience, and that they should be taken into account in decisions regarding technology.[17] But, while this vision of a technological democracy surely needs to be defended, we must also take care that it does not spill over into the analysis and description of our contemporary societies. For by focusing on the premises of technological democracy, social theory is sacrificing the substance for the shadow, and thereby neglecting the study of other spaces, institutions and logics (expert committees, multinationals, venture capital firms) where more decisive developments are happening. It could also be that the political visibility of the new participatory spaces corresponds to a mutation of power, a transformation of government into a governance that takes up the democratic imperative only to instrumentalise it for its own purposes.

Hence the importance for the ecological movement, and for society in general, of having clear historical insight into the past experience

---

16   Amy Dahan-Dalmedico (ed.), *Les Modèles du futur. Changement climatique et scénarios économiques: enjeux politiques et économiques*, Paris: La Découverte, 2007; Aurélien Bernier, *Le Climat otage de la finance*, Paris: Mille et une nuits, 2008.

17   Michel Callon, Pierre Lascoumes and Yannick Barthe, *Agir dans un monde incertain, essai sur la démocratie technique*, Paris: Seuil, 2001.

of technoscience. The challenge in the modern day is perhaps less a matter of defining its exceptional reflexivity, as of considering its past, understanding the success of the devices that produced modern disinhibitions, identifying what still remains of them, and exercising our right to sift through this burdensome legacy. Only in so doing will we be able to emerge from the current strange climate in which we joyously head towards the apocalypse.

Afterword: History as a Smoke Detector

Walter Benjamin spoke of history as a 'fire alarm': the historical premonition of threats should help to 'cut the lighted fuse before the spark reaches the dynamite'. What is the role of history in the face of the Anthropocene? What use is it when there is neither fuse nor dynamite but a generalised conflagration? When prescience has become useless and all that's missing is the courage to act? *Happy Apocalypse* offers no fire hose, no philosophy, no utopia: all it offers is a simple smoke detector. This book is helpful in understanding why contemporary discourse and measures that claim to extinguish the blaze (risk, standards, administrative authorisations, the 'polluter pays' principle, recycling, transition, and so on) fill the air with smoke, rendering it opaque and concealing its source.

In Paris in 2002, when I began the research that led to this book, intellectual debate still stood in the rubble of the Berlin Wall.[1] Despite 11 September, the mood was one of optimism: 'sustainable development' was on everyone's lips, the 'precautionary principle' was fiercely debated and people were enthusiastic about the UN climate summits. The prevailing feeling was one of environmental awakening. Hence the success of the theses capturing this zeitgeist: Ulrich Beck announcing the entry into a 'reflexive modernity', or Bruno Latour – whose work in the sociology of science was nevertheless important for me– predicting 'the end of the modern parenthesis'.

This slightly grandiloquent discourse simply renamed the common idea of 'environmental awareness'. It endorsed the hypothesis that we had

---

1   My research began at the École des hautes études en sciences sociales and the European University Institute, under the direction of Dominique Pestre and Peter Becker.

recently undergone an immense revolution, both social and intellectual, making risk, the environment and the Earth objects of concern and scruples, and that from this revolution all others would flow. The title *Happy Apocalypse* is intended to convey this gratifying and reassuring feeling of sudden revelation.

Reflexivity was a historical, or rather anti-historical, thesis: the past was mobilised only in negative terms in order to highlight our own excellence. On this point, Beck was perfectly explicit. His aim, he wrote, was 'to move the future that is just beginning to *take shape* into view against the *still* predominant past'.[2] My own project was simple: to turn Beck's thesis against itself, to use the categories of *Risk Society* to analyse controversies and struggles that – apart from the case of inoculation – were largely absent from historiography.

One question remained: how had technology imposed itself despite its risks and the awareness of these risks? It was in response to this question that I introduced the idea of 'modern disinhibitions'. Rather than an alleged nineteenth-century faith in progress that we had suddenly lost, it was ad hoc, localised and relatively discrete mechanisms that made technological advance possible: reassuring statistics, convenient nosologies, hygienic theories, standards, liability and insurance regimes, financial compensation, and so on. These were not some grandiose philosophical breaks with the past, or a 'naturalist ontology' (Philippe Descola) specific to the West, but rather little mechanisms, nestled at strategic points of technological modernisation.

The insistence on the smallness of disinhibitions also made it possible to understand this paradox: how could we have forgotten that we had been living in risk societies for two centuries yet claim to be discovering it only now? If reflexivity seemed new, it was because technological modernity had, from its very beginnings, internalised its critiques in a way that was both effective and discrete; it had grown used to dealing with accidents, disasters and struggles; in short, it had had time to hone its disinhibition devices. To now proclaim the novelty of reflexivity was, in essence, to play into the hands of the technological gurus: current modernisation would be made all the more acceptable by the claim that it was now taking place under the anxious vigilance of a society that had become reflexive.

ᘉ

---

2   Ulrich Beck, *Risk Society: Towards a New Modernity*, Thousand Oaks, CA: Sage Publications, 1992, p. 9.

My main thesis on reflexivity has been little challenged. Sociologists and philosophers have retreated into a position of acquiescence, while renewing the idea of a recent rupture. For example, the sociologist Yannick Barthe defended Beck's position in a review by proposing the idea of a gradient of reflexivity.[3] In *Facing Gaia*, Bruno Latour accepted my objection and also pointed out his interest in the notion of disinhibition, while making James Lovelock a new Galileo and the 'irruption of the Earth into politics' a new reflexive epiphany.[4] The success of the notion of the Anthropocene has also revived the illusion of a cosmological revolution. My book *Chaos in the Heavens* (with Fabien Locher) will perhaps dispel this misunderstanding.

Among historians, the idea of a technological modernity achieved with eyes wide open has been taken up by several colleagues with significant modifications. In their work, reflexivity is often thought of as the prerogative of specific groups, precursors, romantics, 'technocritics'.[5] For example, Serge Audier, in his book *La Société écologique et ses ennemis*, marries red and green (in my opinion forcibly) by making 'environmental consciousness' an attribute of nineteenth-century socialist thinkers marginalised by liberals and the productivist left.[6] In general, recent environmental history (following the example of Charles-François Mathis, Rémi Luglia, Caroline Ford and Paul Warde) prefers to consider the question of reflexivity from the angle of a gradual emergence linked to industrialisation and the realisation of its damage, the advance of science, the formation of nation-states managing their resources, or even the Romantic movement.[7] In my book, however, reflexivity is a 'default setting' for societies faced with technological change. The issue is not so much one of 'awareness' – which produces teleological narratives or intellectual genealogies – but rather the opposite: the production of a form of modernising unconsciousness.

---

3   https://laviedesidees.fr/Aux-risques-du-passe.html.

4   Bruno Latour, *Facing Gaia: Eight Lectures on the New Climatic Regime*, Cambridge: Polity, 2017.

5   François Jarrige, *Technocritiques. Du refus des machines à la contestation des technosciences*, Paris: La Découverte, 2014.

6   Serge Audier, *La Société écologique et ses ennemis. Pour une histoire alternative de l'émancipation*, Paris: La Découverte, 2017.

7   Charles-François Mathis, *In Nature We Trust. Les paysages anglais à l'ère industrielle*, Paris: Presses de l'université Paris-Sorbonne, 2010; Rémi Luglia, *Des savants pour protéger la nature. La société d'acclimatation (1854–1960)*, Rennes: PUR, 2015; Caroline Ford, *Natural Interests: The Contest over Environment in Modern France*, Cambridge, MA: Harvard University Press, 2016; Paul Warde, *The Invention of Sustainability: Nature and Destiny, c.1500–1870*, Cambridge: Cambridge University Press, 2018.

My project was therefore closer to 'agnotology', a neologism coined by the historian Robert Proctor to designate the study of the production of ignorance. In my view, these works, interesting and rich as they are – I'm thinking of Robert Proctor's formidable *Golden Holocaust*, which details the strategies of the tobacco industry – give ignorance too important a role and, conversely, are too confident in the power of science and reason.[8] In the nineteenth century, what mattered most for industrialisation was not so much the production of ignorance or false controversies as the manufacture of new knowledge, reassurance devices or social engineering (for example, the confinement of risks to certain populations). It was not so much the silence surrounding technological accidents as safety norms and later occupational insurance. And not so much ignorance of the dangers of the chemical industry as the employment of a lumpenproletariat in these plants and the belief that increased wealth would ultimately improve health. In relation to climate change, it is no longer so much climate scepticism as the revival of processes of disinhibition: carbon credits, sustainable development, green finance, energy transition, and so on.

A third, more pointed debate concerned the true nature of the regulation of industrial pollution in nineteenth-century France. One of the original aspects of my research was the use of civil court archives to address the issue of industrial pollution. Compared with existing work on the subject (André Guillerme, Geneviève Massard-Guilbaud and the work of Thomas Le Roux then in progress), which focused on the actions of the administration, this archival shift led me to emphasise the liberal nature of the new post-1810 regulations.[9] Liberal not in the general sense that it was in favour of industrialists (the ultra-royalists in power in the 1820s were not at all favourable to a clique of entrepreneurs that had emerged in the previous decade), but in the precise sense that these regulations gave a fundamental role to the courts and to compensation for damages. This thesis, which was based on a precise study of the case of industrial chemistry in the early nineteenth century, deserves to be examined in greater depth in other fields. It is an important question, touching on the technical and social effects of the 'polluter pays' principle, its effects

8   Robert Proctor, *Golden Holocaust: Origins of the Cigarette Catastrophe and the Case for Abolition* (Berkeley: University of California Press, 2012).

9   André Guillerme et al., *Dangereux, insalubres et incommodes: Paysages industriels en banlieue parisienne, XIXe-XXe siècle* (Seyssel: Champ Vallon, 2004); Geneviève Massard-Guilbaud, *Histoire de la pollution industrielle en France, 1789–1914* (Paris: EHESS, 2010); Thomas Le Roux, *Le Laboratoire des pollutions industrielles. Paris, 1770–1830* (Paris: Albin Michel, 2011).

on the improvement – or otherwise – of industrial processes and on the construction of environmental segregation.

With hindsight, what would have been worth debating was my basic premise, never made explicit but always present in the background, that the first wave of industrialisation already worsened environmental destruction, that it was basically the first step towards the contemporary abyss. Yet the players at the time perhaps sincerely believed that coal would help save the forests and therefore the climate, that it was in short an 'ecological' solution to the impasses of a wood-based economy. Or if we take the case of soda ash, the Leblanc manufacturing process based on sea salt and sulphuric acid replaced truly staggering quantities of vegetable ash. Alkali plants did indeed produce massive pollution, but it was localised and highly visible because they were close to built-up areas. They certainly damaged crops, but they also relieved pressure on the forests of the Baltic and the kelp fields of Normandy and Spain, and therefore on wild animals and fish. *Happy Apocalypse* is heavily influenced by the history of science and the Foucauldian analysis of power. It is not a truly material history of industrialisation, which largely remains to be written.

Finally, there is another way of reading this book: forget the historiography, the theses and counter-theses, and simply get to know the people who populate its pages. The Norman fishermen who in the 1770s taught the Académiciens des sciences in Paris what kelp was and how young fish lived in it. The alkali workers of Widnes and St Helens who licked the corneas of their acid-sprayed comrades. Mothers and fathers filled with fear – a legitimate fear – at the idea of inoculating or vaccinating their children. And orphaned children, fatherless and motherless, subjected to the sordid experiments of the first vaccinators or carted from village to village to serve as a source of vaccine lymph. There are also the few doubting modernisers: the chemist Clément-Désormes, who (rightly) praised the candle against gas lighting, and Dr Henri-Marie Husson, the man who, more than anyone else, worked to spread vaccination in France and who, in 1803, in his private correspondence, explained that this medical innovation was doomed to failure. When it comes to technology, nothing is ever certain, and this is perhaps the only optimistic conclusion to be drawn from this book. Benjamin also said that history enabled us 'to seize on a memory as it flashes up at a moment of danger'. I hope that *Happy Apocalypse* illuminates some useful memories in these dangerous times.

Paris, October 2019

Bibliography

1. Smallpox inoculation

Blake, John, The Inoculation Controversy in Boston: 1721–1722, *The New England Quarterly*, 1952, vol. 25, pp. 489–506.

Breen, Louise and Cotton Mather, 'The "Angelical Ministry", and Inoculation', *Journal of the History of Medicine and Allied Sciences*, 1991, vol. 46, pp. 333–57.

Herbert, Eugenia, 'Smallpox inoculation in Africa', *The Journal of African History*, 1975, vol. 16, no. 4, pp. 539–9.

Hopkins, Donald, *The Greatest Killer: Smallpox in History*, Chicago: University of Chicago Press, 2002.

Miller, Geneviève, *The Adoption of Smallpox Inoculation, in England and France*, Philadelphia: University of Pennsylvania Press, 1957.

Miller, Perry, *The New England Mind*, Cambridge, MA: Harvard University Press, 1954.

Raymond, Jean-François, *La Querelle de l'inoculation: préhistoire de la vaccination*, Paris: Vrin, 1982.

Razzell, Peter, *The Conquest of Smallpox*, London: Caliban, 1977.

Seth, Catriona, *Les rois aussi en mouraient. Les Lumières en lutte contre la petite vérole*, Paris: Desjonquères, 2008.

Shuttleton, David, *Smallpox and the Literary Imagination, 1660–1820*, Cambridge: Cambridge University Press, 2007.

Van de Wetering, Maxine, 'A reconsideration of the inoculation controversy', *The New England Quarterly*, vol. 58, 1985, pp. 46–67.

2. Smallpox vaccination

Baldwin, Peter, *Contagion and the State*, Cambridge: Cambridge University Press, 1999.

Bercé, Yves-Marie, *Le Chaudron et la lancette*, Paris: Presses de la Renaissance, 1984.

Brunton, Deborah, *The Politics of Vaccination: Practice and Policy in England, Wales, Ireland, and Scotland, 1800-1874*, Rochester, NY: University of Rochester Press, 2008.

Darmon, Pierre, *La Longue Traque de la variole*, Paris: Perrin, 1985.

Hardy, Anne, 'Smallpox in London: factors in the decline of the disease in the nineteenth century', *Medical history*, 1983, vol. 27, pp. 111-38.

Mercer, A. J., 'Smallpox and epidemiological-demographic change in Europe: the role of vaccination', *Population Studies*, 1985, vol. 39, pp. 287-307.

Rusnock, Andrea, 'Catching cowpox: the early spread of smallpox vaccination, 1798-1810', *Bulletin of the History of Medicine*, 2009, vol. 83, pp. 17-36.

Sköld, Peter, 'From inoculation to vaccination: smallpox in Sweden in the eighteenth and nineteenth centuries', *Population studies*, 1996, vol. 50, pp. 247-62.

3. Anti-vaccination movement

Durbach, Nadja, *Bodily Matters: The Anti-Vaccination Movement in England, 1853-1907*, Durham, NC: Duke University Press, 2005.

Faure, Olivier, 'La Vaccination dans la région lyonnaise au début du xix siècle: résistances ou revendications populaires', *Cahiers d'histoire*, 1984, vol. 29, pp. 191-209.

Porter, Dorothy and Roy Porter, 'The politics of prevention: anti-vaccinationism and public health in nineteenth-century England', *Medical History*, 1988, vol. 32, pp. 231-52.

4. Global diffusion of smallpox vaccination

Bulletin of the History of Medicine, 2009, vol. 83.

Arnold, David, *Colonizing the Body: State Medicine and Epidemic Disease in Nineteenth-Century India*, Berkeley: University of California Press, 1993.

Brimnes, Niels, 'Variolation, vaccination and popular resistance in early colonial South India', *Medical History,* 2004, vol. 48, pp. 199–228.

Jannetta, Ann, *The Vaccinators: Smallpox, Medical Knowledge, and the Opening of Japan*, Redwood City, CA: Stanford University Press, 2007.

Moulin, Anne-Marie, 'La Vaccine hors d'Europe, Ombres et lumières d'une victoire', *Bulletin de l'Académie nationale de médecine*, 2001, vol. 185, pp. 785–95.

Perigüell, Emilio Balaguer, *En el nombre de los niños. Real expedición Filantrópica de la vacuna, 1803–1806*, Asociación Española de Pediatría, 2003.

5. History of the body

Baecque, Antoine, *Le Corps de l'histoire, métaphores et politiques, 1770–1800*, Paris: Calmann-Lévy, 1993.

Corbin, Alain, Jean-Jacques Courtine and Georges Vigarello (eds), *Histoire du corps. De la Renaissance aux Lumières*, vol. 1, Paris: Seuil, 2005.

Elias, Norbert, *La Civilisation des mœurs*, Paris: Pocket, 1974.

Foucault, Michel, *Territoire, sécurité, population. Cours au collège de France, 1977–1978*, Paris: Gallimard/Seuil, 2004.

Jahan, Sébastien, *Le Corps des Lumières*, Paris: Belin, 2006.

Laqueur, Thomas, *La Fabrique du sexe*, Paris: Gallimard, 1992.

_____, *Le Sexe en solitaire*, Paris: Gallimard, 2005.

Outram, Dorinda, *The Body and the French Revolution: Sex, Class and Political Culture*, New Haven: Yale University Press, 1989.

Pick, Daniel, *Faces of Degeneration: A European Disorder (c. 1848–1918)*, Cambridge: Cambridge University Press, 1989.

6. History of human experimentation

Bonah, Christian, *L'Expérimentation humaine. Discours et pratiques en France, 1900–1940*, Paris: Les Belles Lettres, 2007.

Chamayou, Grégoire, *Les Corps vils. Expérimenter sur les corps humains aux XVIIIe et XIXe siècles*, Paris: La Découverte, 2008.

Marks, Harry, *La Médecine des preuves. Histoire et anthropologie des essais cliniques (1900–1990)*, Paris: La Découverte, 1999.

Schaffer, Simon, 'Self evidence', *Critical Inquiry*, 1992, vol. 18, pp. 327–62.

Schiebinger, Londa, 'Human experimentation in the eighteenth century: natural boundaries and valid-testing', in Lorraine Daston & Fernando

Vidal (eds), *The Moral Order of Nature*, Chicago: University of Chicago Press, 2004, pp. 385–408.

7. History of medicine, eighteenth century

Barras, Vincent and Philip Rieder, 'Corps et subjectivité à l'époque des Lumières', *Dix-huitième siècle*, 2005, vol. 37, pp. 211–33.

Brockliss, Laurence and Colin Jones, *The Medical World of Early Modern France*, Oxford: Oxford University Press, 1997.

Jewson, Nicholas, 'Medical knowledge and the patronage system in eighteenth-century England', *Sociology*, 1974, vol. 8, pp. 369–85.

_____, 'The disappearance of the sick man from medical cosmology, 1770–1870', *Sociology*, 1976, vol. 10, pp. 225–44.

Jones, Colin, 'The great chain of buying: medical advertisement, the bourgeois public sphere, and the origins of the French revolution', *The American Historical Review*, 1996, vol. 101, pp. 13–40.

Pilloud, Séverine, 'Consulter par lettre au XVIIIe siècle', *Gesnerus*, 2004, vol. 61, pp. 232–53.

Porter, Dorothy and Roy Porter, *Patient's Progress: Doctors and Doctoring in Eighteenth-Century England*, Cambridge: Polity Press, 1989.

Porter, Roy, 'The patient's view: doing medical history from below', *Theory and Society*, 1985, vol. 14, pp. 175–98.

_____, *Patients and Practitioners: Lay Perceptions of Medicine in Pre-industrial Society*, Cambridge: Cambridge University Press, 1986.

Rosenberg, Charles, *Explaining Epidemics and Other studies in the History of Medicine*, Cambridge: Cambridge University Press, 1992.

Vila, Anne, *Enlightenment and Pathology: Sensibility in the Literature and Medicine of Eighteenth-Century France*, Baltimore: John Hopkins University Press, 1998.

Wild, Wayne, *Medicine-by-post: The Changing Voice of Illness in Eighteenth-Century British Consultations Letters and Literature*, Amsterdam: Rodopi, 2006.

8. History of medicine, nineteenth century

Ackerknecht, Erwin, *La Médecine hospitalière à Paris*, Paris: Payot, 1986.

Faure, Olivier, *Histoire sociale de la médecine, XVIIIe–XXe siècles*, Paris: Economica, 1994.

Foucault, Michel, *Naissance de la clinique. Une archéologie du savoir médical*, Paris: PUF, 1963.

Foucault, Michel (ed.), *Les Machines à guérir. Aux origines de l'hôpital moderne*, Paris: Mardaga, 1979.

Gaudillière, Jean-Paul, *La Médecine et les sciences, XIXe–XXe siècles*, Paris: La Découverte, 2006.

Léonard, Jacques, *La Médecine entre les savoirs et les pouvoirs, histoire intellectuelle et politique de la médecine française au XIXe siècle*, Paris: Aubier, 1981.

Maulitz, Russell, *Morbid Appearances: The Anatomy Pathology in the Early Nineteenth Century*, Cambridge: Cambridge University Press, 1987.

Ramsey, Matthew, *Professional and Popular Medicine in France, 1770–1830*, Cambridge: Cambridge University Press, 1988.

Simon, Jonathan, *Chemistry, Pharmacy and Revolution in France, 1777–1809*, Aldershot: Ashgate, 2005.

Weiner, Dora, *The Citizen-Patient in Revolutionary and Imperial Paris*, Baltimore: John Hopkins University Press, 2002.

Weisz, George, *The Medical Mandarins: The French Academy of Medicine in the Nineteenth and Early Twentieth Centuries*, Oxford: Oxford University Press, 1995.

_____, *Divide and Conquer: A Comparative History of Medical Specialization*, Oxford: Oxford University Press, 2005.

9. History of statistics

Bourguet, Marie-Noëlle, *Déchiffrer la France. La statistique départementale à l'époque napoléonienne*, Paris: Archives contemporaines, 1989.

Coumet, Ernest, 'La théorie du hasard est-elle née par hasard?', 1970, *Annales*, vol. 25, pp. 574–98.

Daston, Lorraine, *Classical Probability in the Enlightenment*, Princeton: Princeton University Press, 1988.

Desrosières, Alain, *La Politique des grands nombres. Histoire de la raison statistique*, Paris: La Découverte, 1993.

Hacking, Ian, *The Emergence of Probability*, Cambridge: Cambridge University Press, 1975.

Kavanagh, Thomas, *Enlightenment and the Shadows of Chance: The Novel and the Culture of Gambling in Eighteenth-Century France*, Baltimore: John Hopkins University Press, 1993.

Lebras, Hervé, *Naissance de la mortalité. L'origine politique de la démographie et de la statistique*, Paris: Gallimard/Seuil, 2000.

Rohrbasser, Jean-Marc, *Dieu, l'ordre et le nombre*, Paris: PUF, 2001.

Rusnock, Andrea, *Vital Account: Quantifying Health and Population in Eighteenth-Century England and France*, Cambridge: Cambridge University Press, 2002.

10. History of Neo-Hippocratism

Arnold, David, *The Problem of Nature: Environment and Culture in Historical Perspective*, New York: Wiley, 1996.

Barles, Sabine, *La Ville délétère. Médecins et ingénieurs dans l'espace urbain, XVIIIe–XIXe siècle*, Seyssel: Champ Vallon, 1999.

Cantor, David (ed.), *Reinventing Hippocrates*, Burlington: Ashgate, 2002.

Glacken, Clarence, *Traces on the Rhodian Shore: Nature and Culture in Western Thought from Ancient Times to the End of the Eighteenth Century*, Berkeley: University of California Press, 1967.

Grove, Richard, *Green Imperialism, Colonial Expansion, Tropical Island Edens and the Origins of Environmentalism, 1600–1800*, Cambridge: Cambridge University Press, 1999.

Riley, James, *The Eighteenth-Century Campaign to Avoid Disease*, London: Macmillan, 1987.

11. History of urban police

Carvais, Robert, *La Chambre royale des Bâtiments. Juridiction professionnelle et droit de la construction à Paris sous l'Ancien Régime*, law dissertation, Paris II, 2001.

Milliot, Vincent, 'Qu'est-ce qu'une police éclairée? La police améliioratrice selon Jean-Charles Pierre Lenoir, lieutenant général à Paris (1775–1785)', *Dix-huitième siècle*, 2005, vol. 37, pp. 117–30.

Napoli, Paolo, *La Naissance de la police moderne. Pouvoir, normes, société*, Paris: La Découverte, 2003.

Piasenza, Paolo, 'Juges, lieutenants de police et bourgeois à Paris au XVIIe et XVIIIe siècles', *Annales*, 1990, vol. 45, no. 5, pp. 1189–215.

Williams, Alan, *The Police of Paris, 1718–1789*, Baton Rouge: Louisiana State University, 1979.

12. History of chemical industry, early nineteenth century

Bensaude-Vincent, Bernadette and Isabelle Stengers, *Histoire de la chimie*, Paris: La Découverte, 1992.

Chassagne, Serge, *Le Coton et ses patrons*, Paris: EHESS, 1991.

Crosland, Maurice, *Under Control: The French Académie of Sciences (1795–1914)*, Cambridge: Cambridge University Press, 1992.

Daumalin, Xavier, *Du sel au pétrole*, Marseille: Tacussel, 2003.

Dhombres, Jean and Nicole Dhombres, *Naissance d'un nouveau pouvoir. Sciences et savants en France, 1793–1824*, Paris: Payot, 1989.

Guerlac Henri, *The Crucial Year: The Background and Origin of His First Experiments in Combustion in 1772*, Ithaca: Cornell University Press, 1961.

Minard, Philippe, *La Fortune du colbertisme. État et industrie dans la France des Lumières*, Paris: Fayard, 1998.

Smith, John Graham, *The Origins and Early Development of the Heavy Chemical Industry in France*, Oxford: Oxford University Press, 1979.

Vérin, Hélène, *Entrepreneurs, entreprises. Histoire d'une idée*, Paris: PUF, 1982.

Woronoff, Denis, *Histoire de l'industrie en France*, Paris: Seuil, 1994.

13. History of pollution, France

Baud, Jean-Pierre, 'Le voisin protecteur de l'environnement', *Revue juridique de l'environnement*, 1978, vol. 1, pp. 16–33.

———, 'Les hygiénistes face aux nuisances industrielles dans la première moitié du XIXe siècle', *Revue juridique de l'environnement*, 1981, vol. 3, pp. 205–20.

Corbin, Alain, 'L'opinion et la politique face aux nuisances industrielles dans la ville préhaussmannienne', *Le Temps, le désir et l'horreur. Essais sur le XIXe siècle*, Paris: Flammarion, 1991.

Fromageau, Jérôme, *La Police de la pollution à Paris de 1666 à 1789*, law dissertation, 1989, Paris II.

———, 'La Révolution française et le droit de la pollution', in Andrée Corvol (ed.), *La Nature en Révolution, 1750–1800*, Paris: L'Harmattan, 1993, pp. 59–67.

Guillerme André, Anne-Cécile Lefort and Gérard Jigaudon, *Dangereux, insalubres et incommodes. Paysages industriels en banlieue parisienne, XIXe–XXe siècles*, Seyssel: Champ Vallon, 2004.

Graber, Frédéric, 'La qualité de l'eau à Paris, 1760–1820', *Entreprises et histoire*, 2008, vol. 50, pp. 119–33.

Le Roux, Thomas, *Le Laboratoire des pollutions industrielles. Paris, 1770–1830*, Paris: Albin Michel, 2011.

Massard-Guilbaud, Geneviève, *Histoire de la pollution industrielle en France, 1789–1914*, Paris: EHESS, 2010.

Reynard, Pierre-Claude, 'Public order and privilege: the eighteenth-century roots of environmental regulation', *Technology and Culture*, 2002, vol. 43, no. 1, pp. 1–28.

14. History of pollution, Great Britain

Ashby, Eric and Mary Anderson, *The Politics of Clean Air*, Oxford: Oxford University Press, 1981.

Dingle A., 'The monster nuisance of all: landowners, alkali manufacturers, and air pollution, 1828–1864', *Economic History Review*, 1982, vol. 25, no. 4, pp. 529–48.

Hawes, Richard, 'The control of alkali pollution in St. Helens, 1862–1890', *Environment and History*, vol. 1, 1995, no. 2, pp. 159–71.

Luckin, Bill, *Pollution and Control: A Social History of the Thames in the Nineteenth Century*, Bristol: Hilder, 1986.

MacLaren, John, 'Nuisance law and the industrial revolution: some lessons from social history', *Oxford Journal of Legal Studies*, 1983, vol. 3, no. 2, pp. 155–221.

MacLeod, Roy, 'The alkali acts administration, 1863–1884: the emergence of the civil scientist', *Victorian Studies*, 1965, vol. 9, pp. 85–112.

Mathis, Charles-François, *In nature we trust. Les Paysages anglais à l'ère industrielle*, Paris: PUPS, 2010.

Mosley, Stephen, *The Chimney of the World: A History of Smoke Pollution in Victorian and Edwardian Manchester*, Cambridge: White Horse Press, 2001.

Rosen, Christine, 'Knowing industrial pollution: nuisance law and the power of tradition in a time of rapid economic change, 1840–1864', *Environmental history*, 2003, vol. 8, no. 4, pp. 565–97.

Thorsheim, Peter, *Inventing Pollution: Coal, Smoke and Culture in Britain since 1800*, Athens: Ohio University Press, 2006.

15. History of hygiene

Bartrip, Peter, *The Home Office and the Dangerous Trades: Regulating Occupational Disease in Victorian and Edwardian Britain*, Amsterdam: Rodopi, 2002.

Coleman, William, *Death Is a Social Disease. Public Health and Political Economy in Early Industrial France*, Madison: University of Wisconsin Press, 1982.

Corbin, Alain, *Le Miasme et la jonquille, l'odorat et l'imaginaire social*, Paris: Aubier, 1982.

Hamlin, Christopher, 'Providence and putrefaction: victorian sanitarians and the natural theology of health and disease', *Victorian Studies*, 1985, vol. 28, no. 3, pp. 381–411.

_____, *Public Health and Social Justice in the Age of Chadwick: Britain, 1800–1854*, Cambridge: Cambridge University Press, 1998.

Jorland, Gérard, *Une société à soigner, hygiène et salubrité publiques en France au XIXe siècle*, Paris: Gallimard, 2010.

La Berge, Ann, *Mission and Method: The Early Nineteenth-Century French Public Health Movement*, Cambridge: Cambridge University Press, 1992.

Moriceau, Caroline, *Les Douleurs de l'industrie. L'hygiénisme industriel en France, 1860–1914*, Paris: EHESS, 2009.

Murard, Lion and Patrick Zylberman, *L'Hygiène dans la République. La santé publique ou l'utopie contrariée, 1870–1918*, Paris: Fayard, 1986.

16. History of work

Alder, Ken, 'Making things the same: representation, tolerance and the end of the ancient regime in France', *Social Studies of Science*, 1998, vol. 28, pp. 499–545.

Guillerme, Jacques, and Jan Sebestik, 'Les Commencements de la technologie', *Thalès*, vol. 12, 1966, pp. 1–72.

Jarrige, François, *Au temps des tueuses de bras. Les bris de machines à l'aube de l'ère industrielle (1780–1860)*, Rennes: Presses universitaires de Rennes, 2009.

Kaplan, Steven (ed.), *Work in France: Representations, Meaning, Organization and Practice*, Ithaca: Cornell University Press, 1986.

Kranakis, Eda, *Constructing a Bridge: An Exploration of Engineering Culture in Nineteenth-Century France and America*, Cambridge: MIT Press, 1997.

Pannabecker, John R., 'Representing Mechanical Arts in Diderot's Encyclopédie', *Technology and Culture,* 1998, vol. 39, no. 1, pp. 33–73.

Schaffer, Simon, 'Enlightened Automata', in *The Sciences in Enlightened Europe,* Chicago: University of Chicago Press, pp. 126–65.

Sewell, William, *Gens de métiers et révolutions. Le langage du travail de l'Ancien Régime à 1848,* Paris: Aubier, 1983.

Sonenscher, Michael, *Work and Wages: Natural Law, Politics and the Eighteenth-Century French Trades,* Cambridge: Cambridge University Press, 1989.

Vérin, Hélène, *La Gloire des ingénieurs,* Paris: Albin Michel, 1998.

Wise, Norton M. (ed.), *The Values of Precision,* Princeton: Princeton University Press, 1995.

17. History of gas lighting

Delattre, Simone, *Les Douze Heures noires. La nuit à Paris au XIXe siècle,* Paris: Albin Michel, 2000.

Falkus, M., 'The Early Development of the British Gas Industry, 1790–1815', *The Economic History Review,* 1982, vol. 35, pp. 217–34.

Hunt, Charles, *A History of the Introduction of Gas Lighting,* London: Walter King, 1907.

Matthews, Derek, 'Laissez-faire and the London gas industry in the nineteenth century: another look', *The Economic History Review,* 1986, vol. 39, pp. 244–63.

Nead, Lynda, *Victorian Babylon, People, Streets and Images in Nineteenth-Century London,* New Haven: Yale University Press, 2000.

Schivelbuch, Wolfgang, *La Nuit désenchantée. À propos de l'histoire de l'éclairage artificiel au XIXe siècle,* Paris: Gallimard, 1993.

Tomory, Leslie, *Progressive Enlightenment: The Origins of the Gaslight Industry, 1780–1820,* dissertation, Toronto University, 2010.

Williot, Jean-Pierre, *Naissance d'un service public, le gaz à Paris,* Paris: Rive Droite, 1999.

18. Industrial risk, nineteenth century

Aldrich, Mark, *Death Rode the Rails: American Railroad Accidents and Safety, 1828–1965,* Baltimore: John Hopkins University Press, 2006.

Bartrip, Peter, 'The state and the steam boiler in nineteenth-century Britain', *International Review of Social History,* 1980, vol. 25, pp. 77–105.

Bartrip, Peter and Sandra Burman, *The Wounded Soldiers of Industry: Industrial Compensation Policy, 1833–1897*, Oxford: Oxford University Press, 1983.

Brockmann, John, *Exploding Steamboats, Senate Debates, and Technical Reports*, Amityville: Baywood, 2002.

Bronstein, Jamie, *Caught in the Machinery Workplace Accidents and Injured Workers in Nineteenth-Century Britain*, Stanford, CA: Stanford University Press, 2007.

Burke, John, 'Bursting boilers and the federal power', *Technology and Culture*, 1966, vol. 7, pp. 1–23.

Caron, François, *Histoire des chemins de fer en France, 1740–1883*, vol. 1, Paris, Fayard, 1997.

Chapuis, Christine, 'Risque et sécurité des machines à vapeur au XIXe siècle', 1982, *Culture technique*, no. 11, pp. 203–17.

Defert, Daniel, 'Popular Life and Insurantial Technology', in Burchell, Gordon (ed.), *The Foucault Effect: Studies in Governmentality*, London: Harvester, 1991, pp. 211–33.

Donzelot, Jacques, *L'Invention du social. Essai sur le déclin des passions politiques*, Paris: Seuil, 1994.

Ewald, François, *L'État providence. Une histoire de la responsabilité*, Paris: Grasset, 1986.

Payen, Jacques, *Technologie de l'énergie vapeur dans la première moitié du XIXe siècle*, Paris: CTHS, 1985.

Schivelbusch, Wolfgang, *The Railway Journey: The Industrialization of Time and Space in the Nineteenth Century*, Berkeley: University of California Press, 1986.

Stone, Judith, *The Search for Social Peace: Reform and Legislation in France, 1890–1914*, Albany: Suny Press, 1985.

19. History of proof and objectivity

Collins, Harry, *Changing Order: Replication and Induction in Scientific Practice*, Chicago: University of Chicago Press, 1985.

Daston, Lorraine and Peter Galison, *Objectivity*, New York: Zone Books, 2007.

Pestre, Dominique, *Introduction aux science studies*, Paris: La Découverte, 2006.

Porter, Theodore, *Trust in Numbers: The Pursuit of Objectivity in Science and Public Life*, Princeton: Princeton University Press, 1995.

Schaffer, Simon and Steven Shapin, *Le Léviathan et la pompe à air*, Paris: La Découverte, 1993.

Shapin, Steven, *A Social History of Truth*, Chicago: University of Chicago Press, 1994.

Index